Architectural Design Strategies for Saving Energy in Buildings

Ana-Maria Dabija

Architectural Design Strategies for Saving Energy in Buildings

An Architect's View

 Springer

Ana-Maria Dabija
Commission of Renewable Energies of the Romanian Academy
Technical Sciences Academy of Romania
Bucharest, Romania

ISBN 978-3-031-73540-0   ISBN 978-3-031-73541-7   (eBook)
https://doi.org/10.1007/978-3-031-73541-7

This Springer imprint is published by the registered company Springer Nature Switzerland AG
The registered company address is: Gewerbestrasse 11, 6330 Cham, Switzerland

If disposing of this product, please recycle the paper.

This book is dedicated to my students, to my mentors, to my friends, and to my parents: without their support none of these thoughts of mine would have ever seen the daylight.

About the Author

Prof. Dr. Ana-Maria Dabija has been an architect since 1986. She worked in a design institute and a private company before pursuing a university career with the "Ion Mincu" University of Architecture and Urbanism in Bucharest, Romania. She received her Doctoral Degree in 2000, and has been a doctoral tutor since 2008. Until her retirement, in 2023, Dr. Dabija held several positions in the hierarchy of the "Ion Mincu" University: Vice-rector for innovation and research, Director of the Center of Architectural and Urban Studies, Director of the Architecture Doctoral School of the Ion Mincu University. For over 30 years her scientific interest was on construction principles, sustainable architecture, the relation between architectural design and energy efficiency in buildings. Among the courses developed by Ana-Maria Dabija are "Architectural Detailing Principles," "Contemporary Technological Products and Sub-Assemblies," and "Mistakes in Design – Execution – Use." She is the editor of *Energy Efficient Building Design* (Springer, 2020), and author of *Alternative Envelope Components for Energy-Efficient Buildings* (Springer 2021). Other books worth mentioning (written in Romanian) are *Performant Façade Systems: The Opaque Component, Elements for Stair Design* (also translated in English), *Elements for Designing Windows and Doors*, *Degradation of the Building Envelope*, and *Photovoltaic Systems in Architecture*. Dr. Dabija participated in over 60 international conferences (with published papers) and coordinated technical regulations and scientific research. She is member of the Renewable Commission of the Romanian Academy, corresponding member of the Academy of Technical Sciences from Romania and –, member of the Romanian Union of Architects and of numerous national and international professional and scientific bodies and technical committees.

The future is a door, the past is the key.
Victor Hugo (French writer (1802–1885))

As an architect you design for the present, with an awareness of the past,
for a future which is essentially unknown.
Sir Norman Foster (English architect and designer)

Acknowledgments

Once again my thoughts and thanks go to...

...senior editor *Michael McCabe* who, for the third time, supported me, understood me, and helped me assemble my ideas in the pages of a book,

... the entire Springer Nature team, who patiently carried out the process of production,

...my dearest friends Prof. Em. PhD. Arch *Fani Vavili* and Prof. PhD. Arch. *Paola De Joanna* who kindly accepted to review the book and to whom I am grateful for the suggestions and comments

...my friends who were there for me, listened to my doubts and thoughts, and permanently encouraged me

... and to my unknown reviewers.

About the Book

From the architectural perspective, defined equally as art and engineering science (by corroborating both the philosophical and cultural components with the technical, material ones), the buildings where energy consumption is lower, as a result of integrated measures—both passive and active—compared to the traditional construction systems, could be classified as nZEB.

However, nZEB is a ... diffuse concept: it gives a qualitative direction—"to be very efficient" and requires that this performance is ensured by using unconventional—but numerically quantified energy resources. The simplistic variant is to provide a thermal insulated "blanket" over the envelope and some equipment that produces energy from unconventional sources, which can be calculated according to the regulations.

Adopting such a plain approach, the magic of Architecture would be lost.

All the examples of famous buildings have a common denominator: in the process of creating a spectacular architecture, iconic from the scale of the object to the scale of the city and from the scale of the space to the scale of the detail, the issue of reducing energy consumption is not a target in itself but a consequence of the process of design—by mastering the principles of building physics, the volume of the building, and the spatial configuration—as it is related to the costs of operating the building.

Alternative Envelope Components for Energy Efficient Buildings, published by Springer Nature in 2021, presented the sun and the vegetation as key actors in saving and/or producing energy. *Architectural Design Strategies for Saving Energy in Buildings* can be considered a sequel, traversing the built space through natural ventilation and tackling the aspects of circular economy as leverages of managing the use of resources and energy.

Although active strategies are also mentioned and exemplified, the book focuses mainly on passive design strategies, in a historic perspective, as these are the direct tools of the architect and the most powerful means to save energy and resources, in buildings.

Mies van der Rohe's definition of architecture as *"the will of an epoch translated into space"* is as valid as possible today, the symbolic buildings of our period being designed and modeled in partnership with both technology and Nature, to obtain maximum performance and expressivity.

And with a little luck, these innovative designs that take shape with the integration of innovative materials, systems, and equipment represent the built heritage that will give the measure of the era we live in.

Contents

Chapter 1
Introduction

Why another book on energy efficiency in buildings?

A lot has been written on this topic that the dependence on fossil energy sources—or on any energy source—represents a challenge and a vulnerability for carrying out all human activities. So what does this book bring new?

Maybe the architects' perspective that the primary scope of a building is to ensure the appropriate space for "dwelling"[1]: "'Existential foothold' and 'dwelling' are synonyms, and 'dwelling', in an existential sense, is the purpose of architecture. Man dwells when he can orientate himself within and identify himself with an environment, or, in short, when he experiences the environment as meaningful. Dwelling therefore implies something more than 'shelter.'"

The building can also be a plant for saving or producing energy, and this is an opportunity too good to miss; thus, in the past decades, it "gained" a new physical function that refers to its capacity of saving or/and producing energy, beside the millenary, traditional ones: protection against weathering (wind, water, noise), fire safety, indoor light and air quality, and visual quality of the indoor and outdoor environment. Vitruvius[2] summarized the definition of architecture in only three words, two thousand years ago: Firmitas, Utilitas et Venustas (strength, utility, and beauty).

Other, less concise, more romantic but nevertheless correct definitions were given by great architects of the twentieth century:

Yoshio Taniguchi: *"Architecture is basically a container of something. I hope they will enjoy not so much the teacup, but the tea."*

---

[1] As defined by Martin Heidegger and adopted by Christian Norberg-Schultz in the Preface of his 1979 book "Towards a Phenomenology of Architecture" [1], page 5.

[2] Marcus Vitruvius Pollio, roman architect and military engineer, whose book(s)—De Architectura, Libri X—written somewhere between 30-20 B.C. is the oldest treatise on architecture theory.

A.-M. Dabija, *Architectural Design Strategies for Saving Energy in Buildings*, https://doi.org/10.1007/978-3-031-73541-7_1

Zaha Hadid: *"Architecture is really about well-being. I think that people want to feel good in a space… On the one hand it's about shelter, but it's also about pleasure."*

Daniel Libeskind: *"Architecture is not based on concrete and steel and the elements of the soil. It's based on wonder."*

Arne Jacobsen: *"If a building becomes architecture, then it is art."*

Mies van der Rohe added the social and economic perspective to the definition of architecture: *"Architecture is the will of an epoch translated into space."* It is probably the most complete, yet down-to-earth definition, as the whole history of humanity can be visualized through the history of architecture.

In architecture as well as in the design and production of any commercial item, technology and fashion work together. Technologies broaden the field of building materials and leads to the evolution and development of new construction techniques, paving the road for new approaches, in an area that is governed by the laws of nature: neither gravity nor the laws of thermodynamics or the mechanics of fluids have been canceled, for instance. Building materials and technologies represent the tangible side of the design process. However, integrating a building into the environment, empowering it with significance, shaping spaces, and creating a biunivocal relation with the surroundings while providing the best conditions for carrying out the designated activities are not an attribute of technology or materials, although it cannot be accomplished without them. It is the "wonder" that Daniel Libeskind[3] was speaking about.

Energy intervenes in all design stages—as it intervenes in everything[4]—from the early sketches to the final detail(s), specifications, systems, and equipment throughout the whole life cycle and up to the after-use of a construction.

Most of our lives are carried out in the built environment and inside buildings. According to NHAPS[5] [2] in the mid-1990s "respondents reported spending an average of 87% of their time in enclosed buildings and about 6% of their time in enclosed vehicles.[6]"

Despite this evidence, buildings are considered "responsible for 40% of our energy consumption and 36% of greenhouse gas emissions, which mainly stem from construction, usage, renovation and demolition" [3]. However, they are not. If there were no buildings and the human activities were to be carried outside, in nature (presuming this could happen) wouldn't buildings consume less? Obviously, they would be "responsible" for a much lower percentage of energy consumption. So why blame it on buildings? In reality *human activity* is responsible for the energy consumption; buildings are a result of the human mind and work. Humans, as specialists as well as building occupants, also hold the material leverages—as passive and active design strategies—to diminish, save, or produce energy throughout the whole life of the construction.

---

[3] Daniel Libeskind, Polish-American architect, born in 1946.

[4] The text "Everyrhing is energy and that's all there is to it" is allegedly attributed to Albert Einstein.

[5] National Human Activity Pattern Survey.

[6] The survey was carried out for the USA and Canada between September 1992 and September 1994.

In the past 30 years, different standards were established, for buildings: nZEB, ZEB, carbon-neutral, and so on. In order to accomplish the performances resulted from calculations, a simplistic approach would be to let technology prevail over architecture: by providing sufficient equipment, the values of temperature, humidity, and light are easily met. But the resultant is not architecture.

The truth is that none of the abovementioned building classifications should be related to technology but to transdisciplinarity; they should all be dealing with building design as a complex endeavor that brings together, at the same table and in the same team, specialists from many more different fields of activity than ever before, as technology left its imprint on the building industry, from the drawing board through simulation and augmented reality to the building materials, construction tools, and equipment. Architectural design is about scientific research and validation of theories not only by theoretical assessments but by real-life monitoring[7] of different parameters.

Some existing, old(er) buildings use less energy than others; some use less energy than new buildings and are healthier and costlier. A good question would be: were buildings conceived and erected without bearing in mind the costs that will be paid during their use? Probably not: the architects who designed them certainly had in mind the operational costs.

Were they independent structures that had nothing to do with the environment? Simple or complicated sculptural shapes and forms that randomly change the landscape? Was the cost in use not an element that the design was supposed to take into consideration? Did we invent, now, the awareness regarding the use of resources?

Hardly!

Looking back in history, one can observe that efforts were made to obtain most from a building, beside its resistance, functionality, and aesthetics: thermal comfort, natural cooling, appropriate indoor light, along with weathering protection, protection against noise, and protection against fire. These goals contributed to the overall quality of the building as a commodity and the building principles that were applied were repeated for centuries, like the variations on a same theme. "If it works, don't touch it" applies in all fields of activity.

Introducing and carrying out numerical simulations and estimations is not a new concept either: at least 80 years ago specialists were dimensioning the walls and deciding upon the building materials that should be used according to the same formulas as they do today. Only the levels for the performance requirements changed. And the scope: to save their owners money while operating them, not to save energy or the environment. The questions were not just "can I afford to build it" but also "can I afford to live in it?"

Buildings that could not fulfill their function anymore, which were not structurally safe anymore, were ugly, or maybe consumed too much energy and their use was too expensive were put down and other buildings took their place. It is the case

---

[7] Which unfortunately did not happen where thermal renovation was concerned, in some European countries in the early 2000s, significant discrepancies—up to 30%—between calculated and measured consumption were noted [4].

of the impressive greenhouses of the nineteenth century (Fig. 1.1) that disappeared after the First World War because the owners could not afford to keep them as they required high maintenance costs (although their importance in the history of building structure is as important as it is unexpected). However, the innovative design of these structures invented by John Claudius Loudon [5] and taken to the next level by a gardener who became an architect—Sir Joseph Paxton[8]—can be considered early examples of industrialization or even forerunners of the most common system of curtain walls.

Fig. 1.1 Sir Ribert Paxton The Great Conservatory at Chatsworth. Finalized in 1841 and demolished in 1920. (Public domain. Source: https://en.wikipedia.org/wiki/Chatsworth_House#/media/File:Chatsworth_-_Great_Conservatory_in_the_19th_century.JPG)

The basic principles of building construction are related to the traditional architecture: no matter where the buildings are, the same materials and the same geoclimatic conditions lead to (nearly) the same results.

The relationship between the climatic constraints and the building volumetry, materials, and details is almost tangible: these houses, built by non-educated artisans that inherit, from one generation to the next, the building techniques on the basis of experience, are safe, comfortable, friendly, and healthy. They are warm in winter and cool in summer. Built according to timeless building principles, they represent the core of what is referred to, today, as passive design strategies. In very simple words, passive systems save energy by configuration of the building in the environment and by constructing it using the principles of nature.

---

[8] The author of Crystal Palace. The building was demolished after a fire, in 1936.

Humans are not the only constructors that build with earth, sun, air, and water. We share the planet with other "constructors"— the humble creatures—who need shelters and build them with a science and an efficiency that is still not completely understood by humans (See Chap. 4: Architectural means and tools for providing indoor air comfort).

How did they do it? Probably with tennacity and stubbornness, as if the survival of the species was at stake. And it was, indeed. Or else the species would not have survived.

As technology evolved in time, better comfort conditions were accomplished by integrating within the building, the appropriate equipment that ran on one or another source of energy. Within these active design strategies, a special role is assumed by the systems that produce energy, instead of only using it (mainly originating in traditional fuels). Some contemporary definitions of "active systems" distinguish between the systems that produce energy from renewable sources and consider only the latter as being active.

The line between passive and active may, however, not be as clear and straight as defined above: if "passive" is what comes for free and "active" is what comes after pressing a button, where will, for instance, the blinds fit in? Especially those mechanically or computer operated?!

Is the metallic mashrabiya of the southern facade of the Institut du Monde Arab /World Arab Institute (Fig. 1.2 left) designed by Jean Nouvel[9] a passive or an active system, considering that the lace of panels of connected diaphragms (Fig. 1.2. right) represents a sun protection of the most exposed facade but in the same time, the adjustment of the diaphragms is provided by computer?!

Fig. 1.2 Arch. Jean Nouvel, Institut du Monde Arabe, 1987 South façade and detail of a mashrabiya panel. (Photo left: Fred Romero, CC BY 2.0. Source: https://www.flickr.com/photos/129231073@N06/26805870143, Photo Right: Serge Melki, CC BY 2.0. Source: https://www.flickr.com/photos/sergemelki/3388325271)

---

[9] French architect.

While passive strategies deal with nature's principles and hence the general building configuration is also established taking into consideration the adequate response to the rules of nature, active strategies rely mainly on the type and level of technology and its application in the building industry. As a result, a fair approach would be to integrate the results of scientific discoveries (where applicable) within the building materials and components and use them at their best, replacing some of the traditional products: roof tiles are provided with thin PV films; sun-shading devices have glass integrated PV modules; railings and parapets can be made with thermal-solar panels; phase-changing materials are mixed in plasterboards; and so on. Figure 1.3 illustrates this principle, of building materials that integrate technology: the series of canopies of the Piazza Gae Aulenti in Milan, Italy, designed by architect Cesar Pelli[10] and built in 2012 are provided with solar panels but it is the choice of material and integration in the built space that contributes to the unique personality of the courtyard, inviting the pedestrians to rest and have a coffee at its shadow. The fact that these building materials produce energy is probably the inprint of our society and (hopefully) will be what the future will inherit, as typical, from us. In other words, it is not a goal but a bonus.

Fig. 1.3 Arch. Cesar Pelli, Piazza Gae Aulenti in Milan, Italy. (Photo Ana-Maria Dabija)

Thus, passive concepts as well as active products and systems intertwine in an organic way, complementing each other. Obviously, in the case of active building materials, the architect must comply with the requirements of the specialist (building facilities engineers) for providing the necessary space, routes, and maintenance

---

[10] Argentine-American architect 1926–2019.

routine for the equipment (in these cases, the building material being in the same time a technological equipment).

The driving forces behind active and passive strategies are the same in both cases.

Figure 1.4 proposes a decomposition of passive and active systems to the constituents: earth (for simplification purposes, building materials were framed as "earth" elements, beside natural deposits), sun (with both components of light and warmth), air (including wind, that sets in motion wind turbines), and water.

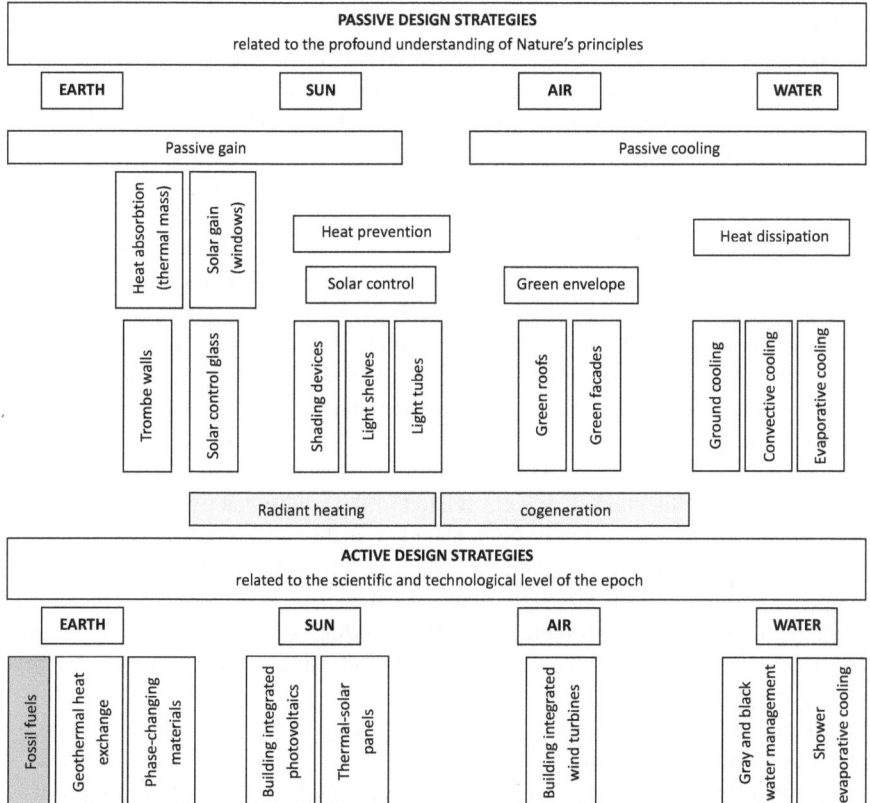

Fig. 1.4 A possible decomposition of passive and active strategies and their relation with the natural environmental agents

Passive gain is accomplished by the thermal mass of the building materials (earth elements) and by managing the solar energy (sun) according to the needs of the moment: maximizing sun gain or providing shadows. As building with the sun was tackled in a previous book,[11] only some aspects (passive systems) were briefly mentioned in the dedicated chapter.

Passive cooling is obtained through air movement and water (see Chap. 4 Architectural means and tools for providing indoor air comfort).

Vegetation[12] contributes as a natural instrument for enhancing the thermal insulation (through the thermal mass of the substrate), the daylight (through the shadowing process), the purity of the air (by filtering the pollutants), and the air humidity (through the process of evapo-transpiration). As quantifying the performances of plants is difficult (so far), using it is a passive qualitative bonus of the building energy balance.

Active means can be tracked down to the same roots: earth, sun, air, and water. The energy resources (fossil fuels) come from the earth, the renewables are related with sun (PV and thermal-solar), air (wind turbines), and water (water management). Hydroelectric power plants were not mentioned in Fig 1.4 as they can be linked with difficulty to the buildings (today; who knows what will tomorrow bring?!).

These systems however are not described or developed in this book as the architects' possibilities to intervene in the technologies are minimal; they should be integrated as innovative building materials (see Figs. 1.1 and 1.2, Chap. 2: Energy and construction policies in the past fifty years and Chap. 5: From yesterday towards tomorrow). Their installation in buildings is the result of knowledge and cooperation between specialists in the effort of creating unique spaces that are also environmental friendly.

Although radiant floors or walls are usually considered "active" approaches, in the traditional architecture they were related to the cooking activity, not indoor air performances. Hence, they may be considered at the border between passive and active strategies, as the hot smoke is, in this case, a by-product that could have been simply exhausted into the atmosphere had it not been directed through a plenum or pipe, in floors or walls. Integrated in the building systems, this "free" source becomes…a passive tool (also see Chap. 4: Architectural means and tools for providing indoor air comfort).

Likewise, the cogeneration plant is in fact a system that produces electricity and—as by-product—steam.

The fact that that steam is used for heating buildings is only one more proof of how brilliant was the inventor.[13] Cogeneration was also not discussed in this book

---

[11] Dabija, A-M., Alternative Envelope Components for Energy-Efficient Buildings, Springer-Nature, 2021, Chap. 4. The Sun – Building Partner of All Times; Passive and Active Approaches, pages 59–88.

[12] The field of architectural means provided by the use of solar energy and vegetation was largely tackled in [6].

[13] Allegedly Thomas Eddison was the inventor.

as, in most cases, it is a strategy addressed to a different level—neighborhood, district, or town.[14]

The toxic gases are mainly an "achievement" of large cities: exhaust gases, gases produced by the operation of the equipment, heat re-reflected by concrete, asphalt, masonry surfaces, and densification (including urban canyons resulting from the spatial configuration of urban areas) are all caused by human activity and contribute to the significant temperature increase in the densely populated city centers and to the occurrence of the urban heat island. "Residual heat" (energy released by people's activities, equipment, industrial processes) also contributes to the raising of the urban heat island index. In other words, not only the industrial activities create excessive heat but also the use of the domestic equipment (air conditioning, for instance, but also other heating and cooling devices) lead to the growth of residual heat in the outdoor space. Changing the source that produces energy does not change the fact that the equipment releases heat in the environment, unless the waste heat is converted into energy that can be reused in other processes (which is indeed an important leverage in reducing the effects of the urban heat island, along with measures like increasing the green/planted areas, the water surfaces, the reflectivity of the surfaces).

Moving from the individual level to the level of the society, the history of mankind can be tracked in the history of architecture: new technologies had an impact on peoples' lives and led to the invention of new building materials in the construction field; periods of economic flourishment led to development of the arts and architecture just as the periods of economic crisis can be identified by analyzing the buildings as systems or by their components.

Architecture is conservative, in the sense that it gives the same response to the action of natural and anthropic agents that act according to the same pattern for thousands of years. What has changed? Our mentalities and some of the means—technologies and, as a consequence, some construction materials.

What buildings will the future keep as representative for our time? Probably those that were—or will be—conceived and built with the respect of the environment and integrating the contemporary technologies within the building materials and systems. Or, in other words, those buildings that emphasize and not deny the effect of the forces of nature, linking in a contemporary result tradition and innovation in design.

We look at buildings in wonder sometimes, in admiration sometimes but we forget that they are the survivors of time and history: some buildings disappeared as a consequence of the occurrence of natural agents or anthropogenic activities. Earthquakes, tsunamis, floods, volcano eruptions, and climate change led to the destruction or/and abandon of settlements; wars, decisions of crowned heads, or political leaders changed the faces of urban settlements; social utopias applied on human categories led to architectural programs that eventually failed.

---

[14] These systems have been used in Romania since the mid-1960s. Working on fossil fuels, the thermal agent was distributed through the urban infrastructure up to the buildings and apartments. The principle is correct but the fuel needs to be re-evaluated.

In other words, most of the built patrimony (yes, the building stock represents a patrimony where philosophy, culture, and engineering are embedded) represents a valuable asset that needs to be understood and the lessons need to be learned; it is not now and we who invented the wheel. Calculations back up solid theories and mathematic models and simulations need to replicate as close as possible the natural conditions of a specific site.

When referring to buildings, energy should not be regarded as a goal in itself but as a consequence of the architectural concept. The interdisciplinary collaboration between specialists from the early(est) stages of design leads to solutions that save energy and natural resources in general. The more detailed and complete is the "script" of operating the building, the more problems are revealed, the less compromises are made in construction and use and the more efficient use of energy is registered. While the passive means are directly in the architects' hands, while the choice of building materials is also an issue where the architect holds the decision, in terms of equipment it is important to choose knowingly, with the support of the partner-specialists. This is why such equipment was not tackled here but is only mentioned in Chap. 5 where case-studies are presented.

In a way of speaking, technologies wrap up old principles, giving new light to millennial building concepts. It is interesting to study the trajectory of a technological system: why, where, and how it was used for the first time (at least what we think was "first"), how it was changed and adapted, and how far away from the original role it is today. This, however, might be the topic of anther book.

References

1. Norberg-Schultz, C. (1979). Genius loci. Towards a phenomenology of architecture. Rizzoli.
2. Klepeis, N. E., Nelson, W. C., Ott, W. R., Robinson, J. P., Tsang, A. M., & Switzer, P. et al. (2001, March 1). *The National Human Activity Pattern Survey (NHAPS) A Resource for assessing exposure to environmental pollutants.* https://digital.library.unt.edu/ark:/67531/metadc719357/m2/1/high_res_d/785282.pdf
3. https://commission.europa.eu/news/focus-energy-efficiency-buildings-2020-02-17_en. Accessed 2022.
4. Sunikka-Blank, M., & Galvin, R. (2012). Introducing the prebound effect: The gap between performance and actual energy consumption. *Building Research & Information, 40*(3), 260–273. https://doi.org/10.1080/09613218.2012.690952
5. https://victorianweb.org/art/architecture/iron/loudon.html
6. Dabija, A.-M. (2021). *Alternative envelope components for energy-efficient buildings.* Springer-Nature Publishing House.

Chapter 2
Energy and Construction Policies in the Past Fifty Years

Introduction

The past century seems to have been rattled by several major events: not only wars—extinct or ongoing—but the evolution of technologies led to modifications of the behavioral pattern and reshaping at the societal and economical scale. The natural changes induced by the generational conflict turned into a deep fracture between generations, as almost everything seems to be questioned and tilted. From the end of the nineteenth century, technologies have made huge leaps and imposed different lifestyles (and altogether new concepts of life). World communications fueled the changes on a larger scale than ever before, with the "+" and "–" that they imply. The traditions were left aside for a while and buildings throughout the world look more or less alike—huge geometric shapes covered with metal and glass—as technology prevails. New materials, new products, and new systems are scattered around the world with little or not at all reminiscence of the *genius loci*.[1]

The story of nZEB, ZEB, and other contemporary building labels that we refer to today seems to begin with the oil crisis of 1973 and 1979, when the energy price rose, amending the way we related to energy and resources. With periods of experimenting that alternate with periods of relaxation or restraints, these 50 years were divided by Dr. Steven Fawkes (in *A brief history of energy efficiency* [1]) in six distinct phases, relevant from the point of view of modern energy management:

Phase 1: 1973–1981—"energy conservation phase"
Phase 2: 1981–1993—"energy management phase";
Phase 3: 1993–2000—"energy procurement phase";
Phase 4: 2000–2010—"carbon reduction phase";
Phase 5: 2010–2020—"energy efficiency phase";
Phase 6: 2020–2030 – "efficiency as a resource phase"

---

[1] Christian Norberg-Schultz, Genius loci. Towards a Phenomenology of Architecture, Rizzoli, 1979.

© The Author(s), under exclusive license to Springer Nature Switzerland AG 2024   11
A.-M. Dabija, *Architectural Design Strategies for Saving Energy in Buildings*,
https://doi.org/10.1007/978-3-031-73541-7_2

Beginning with the 1970s each decade had a specific philosophy of approaching the problematic of energy and the way it had an influence on the economy.

Taking this classification as a potential reference frame and zooming in on these decades, adding other events and actions that took place in each specific period, the general context in which the energy management turned into a preoccupation becomes more colorful, as it was not one but several actions that led to changes in the way buildings were conceived, designed, built, used, and maintained. However, the general quest is not about energy as much as it is about life standard: the echoes of the fear that natural resources might be ending led to the adoption of new design philosophies in the building field.

In the early 1970 (late 1960s) a new architectural concept was launched by the Hungarian architect and Princeton Professor Victor Olgyay: the bioclimatic approach in architecture. His book, *"Design with Climate,"* first published in 1963, is still outstanding as he identifies and studies the leverages and means of the (architect) designer, to maximize the well-being of the occupants by using the environmental agents.

The 1970s

Environmental Policies

The political event that opens the first decade of our analysis—the 1970s—is the United Nations Conference on the Human Environment, held in Stockholm, Sweden, in 1972. It is based on previous hypothesis, research, literature, or actions that can be tracked back some hundreds of years.[2] One of the recommendations of the document that was adopted in this conference, the Stockholm Declaration and Plan of Action, established a connection between energy and environment issues that had to be addressed "in a co-operative spirit by all countries, big and small, on an equal footing [...] to effectively control, prevent, reduce and eliminate adverse environmental effects resulting from activities conducted in all spheres" (Principle 24 [2]).

In the same year, 1972, the Club of Rome presented their first—and most famous—report: The Limits to Growth.[3] Based on previous theories deriving from the eighteenth-century *An Essay on the Principle of Population* written by Thomas Robert Malthus but with the aid of computer modeling (a revolutionary approach, in the early 1970s), the book offers an apocalyptic view of the future where the resources would be depleted and exhausted in less than a century. Although the

---

[2] Legislation In England or in the Republic of Venice was protecting the forest against possible depletion as early as the sixteenth century.

[3] A brief history is represented by the chapter "Principles of Sustainability: History and Evolution in Dabija, A-M., Alternative Envelope Components for Energy-Efficient Buildings, [3], pages 5–28.

predictions were (fortunately) not confirmed, there are supporters and advocates of the theory.

In terms of energy management, during this decade (or Phase 1 in Dr. Steven Fawkes' analysis) the aim to save energy was either by "switching off" the equipment or by developing research to lead to integrating systems that used alternative energy sources.

The oil crisis of 1973 swept away the dust of another ancient technology: geothermal heating. It became popular in Sweden and developed global acceptance [4] in countries that had hot springs and geysers.

Building-Related Diseases: Consequence of the "Switch Off"

In the field of building facilities, a re-evaluation of the necessary outdoor airflow rate for ventilation was established (also as a measure of saving energy to cope with the consequences of the 1973 energy crisis): a limit of 0.14[4] cubic meters/minute/person[5] was set, from a ratio that had been 6 to 12 times higher[6] before the Arab oil embargo. The crisis of 1973 was one of the catalysts that triggered the idea that energy should be used efficiently.

This drastic reduction of the ventilation airflow rate, combined with the emissions of volatile compounds—found in the carpets and furniture—as well as carbon monoxide from the heating devices, seem to have been the cause of poor indoor air quality, resulting in what is known as the *sick building syndrome (SBS)*. The phenomenon was identified in the first years mostly in the case of female workers,[7] presumably explained by the fact that the male working environments were larger, more flexible, and better ventilated, providing better physical and psychological health and security [6].

The term, as we know it today, was coined in 1984, by a biophysicist and today, nearly 50 years after the problem was identified, it covers a range of health issues that are related to the indoor air conditions: chemical and biological contaminants, poor ventilation, and inadequate temperature.

---

[4] 5 cubic feet.

[5] In 1836, a mining engineer, Tredgold calculated that a person needs 0.10 cubic meters/minute (around 4 cubic feet/minute) of clean/unvitiated air to eliminate the CO_2 that was deposited in his lungs, the body moisture, and the candle toxicity. In the years to come, in the late 1800s, the quantitative ventilation ratio of the buildings increased significantly: from 30 cubic feet per minute to 60 cubic feet per minute, calculation recommended by J. Billings, a physician who was mainly concerned with limiting the spreading of tuberculosis [5].

[6] The debate between constructors—architects and engineers—who considered that 30 cubic feet/minute/person was reasonable and the physicians who reccommended 60 cubic feet/minute/person, as they recognized that pacients were healing faster and diseases would spread less in ventilated spaces [5].

[7] And there were debates whether it is real or just hysteria.

The clinical reactions are divided in "dry" symptoms (stuffy nose, dry throat, dry skin), "allergic" symptoms (runny or itchy nose, watery itchy eyes), "asthma" (chest tightness), and arid "general" (undue lethargy and headache) [7].

While most common SBS symptoms disappear in a few hours after the occupant leaves the office and—obviously—over the weekend, some may, however, generate serious illnesses. The most famous case is probably the story of *Legionella pneumophila* a disease caused by a bacterium that infects between 8000 and 18,000 persons annually in the USA [8], with a death rate of 10–15% [9].[8]

Legionella was discovered in 1976, after a tragedy: the Pennsylvania State American Legion celebrated its 5-th annual convention [10]. Four thousand members, with families and friends, were gathered in Philadelphia between July 21 and July 24. The Bellevue-Stratford Hotel hosted 600 of the participants. The day after the convention, a strange form of severe cold-like disease hit the participants, causing 200 victims and, by the end of the week, the first Legionnaire lost his life. By mid-August 221, individuals contacted the mysterious disease and 34 of them died of an unknown form of pneumonia. It took the researchers and investigators 6 months to identify the cause of the disaster. Eventually it proved to be a type of unknown *Bacillus*, sheltered in the hotel air conditioning system and spread through aerosolized mist or vapors. As it was related to the Legionnaires' convention, it was called *Legionella*; and because it attacked the lungs, *pneumophila*.

Further research proved that it lives in cooling systems, cooling towers (in humid ducts), taking advantage of the water management problems, at temperatures that range between 15 °C and 50 °C.[9] The disease is treated with antibiotics but the statistics report that the fatality rate of the Legionnaires' disease can rise to 50% if the antibiotic treatment is delayed.

It took a decade of research, experiments, and validation procedures to establish the means to neutralize the bacteria: heat disinfection, ozonation, UV light, chlorine, or specific biocide exposure [11]. Design and maintenance requirements as well as procedures for intervention were established in the effort of diminishing its occurrence [12].

Legionella proved to have been the cause of other massive illnesses prior to 1976 as well as after, in the USA and in other countries, mainly in hotels and hospitals [10]:

> *"In 1965, 81 patients at St. Elizabeth's psychiatric hospital in Washington, D.C., developed pneumonia and 14 of them died. [...] In July 1968, 144 visitors to, and employees of, the Pontiac, Michigan, Health Department developed a relatively mild illness that was called Pontiac Fever. It was determined, then, that the disease was not caused by any known environmental allergens, toxins or viruses. However, serum from those with Pontiac Fever was also found to contain antibodies to L. pneumophila. The spread of the disease was traced to a leak in the building's air duct that allowed water from the air conditioning system to enter the building's atmosphere. [...] Another epidemic of pneumonia at Philadelphia's Bellevue-Stratford Hotel occurred in 1974, when 20 members of the Independent Order of Odd Fellows developed symptoms one to nine days after attending a meeting in the hotel's main ballroom. Two of those people died. [...] In April 1985, 175 patients were admitted to hospitals in Stafford England with chest infections or pneumonia-like symptoms. A total of 28*

---

[8] The data is for the USA but with the probability of occurrence in other countries.

[9] 59 °F–122 °F.

people died. Medical diagnosis showed that Legionnaires' disease was responsible [...] In
March 1990, an outbreak of Legionnaires' disease at a flower exhibition in the Netherlands
caused 318 people to become ill and at least 32 died. In April, 2000, 4 people died in
Melbourne, Australia out of 125 confirmed cases".

And still counting.

Another building-related disease can be caused by the miss-use of air conditioning: the human body adapts with no difficulty to air temperature changes of less
than 4 °C. Consequently, air conditioning may cause directly or indirectly other
building associated diseases:

– Directly, as the body resents a low indoor temperature (from, for instance, +35 °C
  outdoors to +20 °C indoor) and the effort to adapt represents a major stress that
  can cause illness, from a simple cold to pneumonia, asthma, or other respiratory
  diseases
– Indirectly, as it circulates the air in an indoor environment, raising dust, mold,
  and pollutants.

Sick building syndrome began as a diffuse disease based on inappropriate ventilation (as the drastic reducing of the airflow ventilation rate led to airborne contaminants); nowadays it may occur in indoor environments where the air conditioning
and humidification equipment maintenance are inadequate and where the problem
of providing the appropriate ventilation system persists. Natural ventilation[10] on the
other hand is a good alternative, whenever applicable.

Building Strategies

The 1970s is the period of rediscovering forgotten principles like the Trombe walls[11]
or the (integrated) solar panels[12]; these were instruments which supported the development of a specific architectural trend: the solar architecture. Nevertheless, some
of the instruments that defined the solar architecture of the 1970s were not new
inventions; they existed in the late nineteenth century and were abandoned in favor
of more technically performant and financially efficient systems.

---

[10] More on the topic of natural ventilation in Chap. 4.

[11] The "Trombe wall" was tackled in [3], pages 74–77. In a nutshell, the Trombe wall consists of an
assembly of a heavy wall, a buffer zone, and a glass closing, facing the (sunny) exterior environment; the thick, heavy wall separates the buffer zone from a (main) room; it is provided with
transversal apertures in the lower and upper part, so that the warm air from the buffer zone would
rise to the ceiling and the warm air would pass through the holes into the main room, while the
cooler air from the room, with a higher density, would pass through the holes situated on the lower
side of the wall, into the buffer zone where it would warm up. Hence a permanent convection loop
would be activated. In order to increase the absorption of solar radiation, the surface of the massive
wall facing the buffer zone should be dark.

[12] Solar water heaters were an alternative to electric heating of water when the costs of electricity
peaked, in many parts of the world: USA as early as the nineteenth century, Japan in the 1950s,
Australia in the 1960s and 1970s; after the energy prices fell, the interest in these systems fell as
well, as they were expensive and inefficient in winter time (for instance).

In what solar power is concerned, the 1970s is a flourishing period in the development of the PV technology. Several systems of Building Integrated Solar Thermal Roofing were patented [13], aiming to harness solar energy. Amorphous cells were produced for the first time, with an efficiency of 2.4%.

Solar power began to be used on a larger scale for different industrial applications but due to the poor efficiency—the performance of the PV was as low as 6%—and the very high production costs, punctual and rather small applications were targeted. As the political and economic decision-makers were aware that, at these prices, the solar-cell market could not exist, the price dropped by 80% in 1973, although an even less efficient model was introduced (compared to the standard one of the period) [14, 15].

The solar architecture of the 1970s integrated passive and active design strategies, from adequate orientation of the building and its components to the best valorization of the energy of the sun. This decade marks the beginning of innovation and integration of solar-based systems in architecture as roof integrated systems [16]. One of the most famous experiments was carried out by the Institute of Energy Conversion under the University of Delaware, in 1973: Solar One,[13] a building that ran on a combination of solar-thermal and solar-photovoltaic power. This building is considered to be "the first house to convert sunlight into electricity and heat" (Fig. 2.1) and, in the same time, the first thin-film solar cell laboratory in the world [17].

Fig. 2.1 Cover of the solar one brochure, 1973. (Photo credits: MSS 483, Karl Wolfgang Böer papers, special collections. By courtesy of University of Delaware Library, Newark, Delaware)

---

[13] The team of researchers was conducted by Karl Böer, founder of the Institute of Energy Conversion at the Delaware University.

"Solar One was a two-bedroom, 1300-square-foot house built to demonstrate solar energy's ability to provide both power and heat for a residence and to provide data for further research. Many components — including the solar modules — were built by hand because they were not commercially available in 1973" [17]. As it was opened only few months before the Arab embargo, it soon became an attraction and an example in the state: it was fully operated by the power of the sun, in both electricity and heating. Fifty years later, the house still stands but stripped off its original solar panels.

The 1980s

Considered by Dr. Steven Fawkes the "energy management phase," this period is characterized by the development of software monitoring on a large scale, mainly for bill analysis. Building Energy Management Systems, BEMS (now BMS), were adopted mostly by local authorities, in countries with advanced economies, as these technologies were—as expected—very expensive at first but became more affordable as the prices of the computers fell. Countries with lower economic development did not manage to adopt BMS technologies due to the high costs.

1987 was the year when the European Commission adopted the Single European Act (SEA) with the intention of removing the barriers that hindered the free circulation of goods between the countries that formed the ECSC.[14] This Act triggered the separation—and the economic competition—between the producers, transporters, and distributors of energy, which represented the beginning of the liberalization of the energy sector.

Energy prices began to decline, the process being continued over the following years. The consequence, in the building sector, was that the interest for solar houses was lost toward the end of the decade, as their prices were high compared to their efficiency and the pay-back period.

Environmental Policies

1987 was the year the Brundtland Report—Our Common Future—was presented in the World Commission on Environment and Development, a body established by the UN in 1983 for this purpose: to monitor the environmental and developmental problems and to propose solutions. It was the most visible moment of a trajectory that began several centuries ago and eventually led to the consolidation of the definition of sustainable development as we know—and accept—and to the identification of leverages that may accomplish it at the scale of the planet.

---

[14] In 1950, at Robert Schuman's (the French foreign minister) initiative, a common market for coal and steel was created: ECSC (European Coal and Steel Community). The participant countries were France, Italy, Belgium, the Netherlands, Luxembourg, and West Germany.

Science and Energy

While the building sector experienced a drawback in integrating active design solutions that tackled solar energy systems, other industrial fields continued to experiment the possibilities of using the sun as the source for providing energy: in 1981 *Solar Challenger* (Fig. 2.2), the first human-piloted solar-powered aircraft flew from California airport [18] and, in 1982, "The Quiet Achiever," the first vehicle that used only solar energy was driven from Perth to Sydney [19].

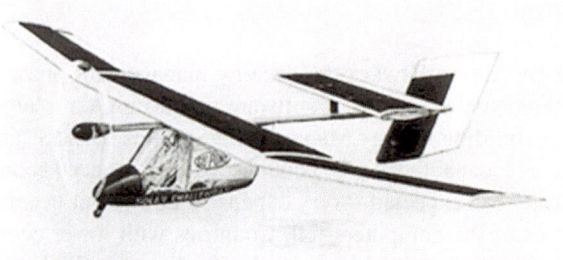

Fig. 2.2 Solar challenger. (Published by NASA unattributed—http://www.nasa.gov/centers/dryden/news/FactSheets/FS-054-DFRC_prt.htm Public Domain. Source: https://en.wikipedia.org/wiki/MacCready_Solar_Challenger#/media/File:Solar_Challenger_drawing.jpg)

The photovoltaic industry developed through innovative research (in 1985 the efficacity of the silicon solar cells reached 20% in laboratory conditions).[15] Facilities that produce energy were built (Fig. 2.3), using concentrating solar collectors[16] as well as solar-thermal sources.[17]

Fig. 2.3 Solar one and solar two power plant, Daggett, California. (Photo work of the United States department of energy—Public Domain. Source: https://en.wikipedia.org/wiki/The_Solar_Project#/media/File:Solar_two.jpg)

---

[15] In the laboratories of the University of South Wales.

[16] Situated in Daggett, California, Solar One was a demonstrative power-tower system [15].

[17] The most famous—and largest—power plant is located at Kramer Junction.

Experiments of integrating photovoltaic arrays in buildings were carried out as well (but the prices seem to have been still high for the middle-class house owners).

The 1990s

It is the decade of political statements, principles, and methods of evaluating the use of energy resources in a worldwide organized approach. United Nations conferences represent the spearhead of this endeavor.

Energy and Environmental Policies

The United Nations Conference on Environment and Development in Rio de Janeiro in 1992, also known as the Earth Summit, established a set of 27 principles for sustainable development among which we mention "the essential task of eradicating poverty as an indispensable requirement," as well as stopping pollution and the depleting of natural resources. The Agenda 21—the document that was adopted at the end of the Summit— shows that an official global, cross-border consensus is required in order to ensure the protection of the environment. This paper is divided into sections that refer to the social- economic dimensions, the conservation and management of resources, as well as the role of major groups; in other words, it creates the foundations on which the theory of sustainable design is constructed: People-Planet-Profit, the pillars of sustainability.

Another outcome of the Earth Summit was the setting up of the United Nations Framework Convention on Climate Change. It entered into force in 1994.

Based on the 1994 United Nations Framework Convention on Climate Change, a new document was elaborated in 1997: the Kyoto protocol, which "operational- izes the United Nations Framework Convention on Climate Change by committing industrialized countries and economies in transition to limit and reduce greenhouse gases (GHG) emissions[18] in accordance with agreed individual targets" [21]. The protocol provided the context in which countries can act to reach the specified goal: to reduce the emission of gases that contribute to global warming. Although less referred to, special sessions of the UN general assemblies on climate change were held every 5 years: Rio + 5 was held in 1997 in New York.

---

[18] According to [20], Roy W. Rising of Valley Village writes in 2011: "Today's report focuses on a bundle of gases that comprise a very small part of total of 'greenhouse' gases. It totally disregards the long-known fact that about 95% of all 'greenhouse' gases is WATER VAPOR! Spending bil- lions of dollars to alter a few components of the 5% won't affect the natural course of climate change."

Great Britain launched Building Research Establishment's Environmental Assessment Method (BREEAM), in 1990, the first assessment method that evaluated and rated the new buildings, based on a grid of "green" criteria, aiming to boost the sustainable design principles from theory to practice; its example was followed in the following years by other organizations and countries that produced their own systems of evaluation: Leadership in Energy and Environmental Design (LEED) in the USA, Green Stars in Australia, Haute Qualité Environnementale (HQE) in France, Bewertungssystem Nachhaltiges Bauen für Bundesgebäude (BNB) in Germany, and probably other less well known.

In 1993 the first Green Building Council was founded, with the mission "to promote sustainability-focused practices in the building and construction industry" and to "bring together all actors across the built environment value chain to advance green building" [22]. In 1999 it was transformed in the World Green Building Council (WGBC). Since then, several countries founded national Green Building Councils, forming a network coordinated by WGBC.

The "energy procurement phase"[19] (1993–2000) was the decade that witnesses the privatization of the utilities (liberalization of the market) and led to a decline of energy prices. Therefore, the energy consultancy market decreased accordingly, and energy management policies were incorporated in wider environmental initiatives, in the hope that the approach would be a holistic, interdisciplinary one, that would improve the efficiency of the measures.

Alternative Applications in Buildings

In the field of solar energy, during the last decade of the twentieth century, research for developing solar cells with higher conversion efficiency continued, in parallel with the construction of larger solar energy plants. The idea of integrating photovoltaic cells within building products and systems led to the invention of flexible solar roof shingles that use amorphous silicon.

An innovative design approach, backed up by the market production, was the insertion of building integrated photovoltaic (BIPV) panels in the façade of a New York skyscraper—the Four Times Square (Fig. 2.4). Although the efficiency of the vertical panels is substantially lower compared to the roof integrated panels, the message sent was of commitment to green or environmentally friendly design.

---

[19] As stated by Dr. Steven Fawkes.

Fig. 2.4 Four Times Square in 2013. (Photo: Eden, Janine and Tim, CC BY 2.0 DEED. Source: https://www.flickr.com/photos/edenpictures/9002148431)

The BIPV facade represented a turning point as it launched a new building product for the envelope (that requires higher qualification in the installation) and added a new function[20] to the building envelope: provider of energy.

The building was presented by Suzanne Stephens in the prestigious magazine Architectural Review, in 2000 [23]:

> *"At Four Times Square, alternative-energy fuel cells, which convert natural gas into electricity without combustion, were installed to generate electricity on-site. But because of their size, their expense, and the fact that they run continuously, only two 200-kilowatt fuel cells, instead of the desired eight, were purchased, covering just 8 percent of the building's electricity. Photovoltaic (PV) cells constituted another method of generating power on-*

---

[20] The traditional functions of the building envelope are to protect the occupants against weathering (rain, wind, snow), sun, fire, and noise and to provide light and ventilation (where applicable).

site. Two hundred eighty-eight PV panels replaced spandrels on 2955 square feet, or half
of a percent, of the building's walls. The electricity created would supply five or six houses
but meets only about a half of a percent of the building's energy needs."

However, according to the press of the time [24], the interest for solar systems of producing energy massively declined by 1992: "In the 70's and 80's Long Island homeowners embraced solar energy as the alternative of the future. They bought more than 10,000 systems, ranging from equipment that could heat and energize a large house to smaller units for hot water or to warm up a swimming pool. Now such purchases are as frequent as blizzards, and interest in this energy alternative seems to have evaporated. [...] This is a shame, because in the last decade the technology has made great advances. One lesson we should have learned from the gulf war was that we need to keep our options open on energy dependence."

Forgotten for almost 50 years, the vegetated roofs and facades reappear in the built landscape, as punctual approaches or as... statements.[21] The Fukuoka ACROS building in Japan proves once again the commitment of architect Emilio Ambasz—indeed "a pioneer of the relationship between buildings and nature"[22]—to bring back the nature in the heart of the city; in this case, to continue the city park on the terraces of an outstanding construction. Almost 30 years after its inauguration, the planted terraced conference hall still fulfills its function of a green lung for the city (Fig. 2.5).

Although his first vertical planted wall was built in the late eighties, Patrick Blanc's vertical gardens (Fig. 2.6) began to flourish (so to say) in the 1990s and in the first decade of the twenty-first century.

---

[21] More on the topic of vegetated roofs and facades in Dabija, A-M., Alternative Envelope Components for Energy-Efficient Buildings Chap. 3 The Living Envelope of the Buildings: History and Evolution, pages 29–58 [3].

[22] As stated in 2020 with the occasion of awarding the architect, for the fourth time, the Compasso d'Oro, prize [25].

Fig. 2.5 The ACROS Building in Fukuoka, Japan. Architect Emilio Ambasz. (Photos: above Sung Yong Woo, CC BY-ND 2.0 DEED. Source: https://www.flickr.com/photos/12273759@N04/7439732036; below Kenta Mabuchi, CC BY-SA 2.0 DEED. Source: https://www.flickr.com/photos/kentamabuchi/5920306109)

Planted roofs, terraces, or walls do not only represent a way of bringing nature back into the cities but they provide a healthier environment for all the living creatures, humans included, and restore the characteristics of the original microclimate.

Fig. 2.6 Fondation
Cartier, Architect Jean
Nouvel, Botanist Patrick
Blanc, 1998. (Photo Rory
Hyde, CC BY-SA 2.0
DEED. Source: https://
www.flickr.com/photos/
roryrory/2520086695)

Around the same time, in 1997 the book *Biomimicry: Innovation Inspired by Nature* written by Janine M. Benyus was the starting point of a new architectural direction. The book does not refer to buildings but the idea that the constructions can—or should—replicate principles of the natural world began to take shape. In fact, the concept was not a new one (it had not yet been theorized), considering that in the late nineteenth century, Antonio Gaudi, the fabulous Catalan architect, was designing in Barcelona buildings and structures inspired by the surrounding nature (Fig. 2.7).

Fig. 2.7 Structures in the
Park Güell. Architect
Antonio Gaudi,
1900–1914. (Photo
Ana-Maria Dabija)

The Twenty-First Century: The Two Thousands

Political Approaches in the Field of Energy and Environment

During this decade many countries adopted measures—by tax incentives—to boost
the installation of solar systems (thermal or photovoltaic). Hence, in the USA an
average annual growth rate of 50 percent was reported by Solar Energy Industries
Association as well as a drop of 70% of the price of installation [16].

With the turn of the century, a new agenda made its way to the general attention:
climate change. The quantifiable result would be to monitor the carbon emissions.
Dr. Steven Fawkes[23] considered this period as Phase 4—"carbon reduction phase."

---

[23] In A brief history of energy efficiency [1].

One of the most well-known events of the decade was Al Gore's[24] *An Inconvenient Truth*, the American documentary film released in 2006 that represented the basis of his conferences regarding the anthropogenic causes for global warming.

About the same time a "carbon market" was established that began to monitor and penalize sectors with high carbon dioxide emissions.

In the field of sustainable development, the decade withstood another world summit at the highest level of participation: Johannesburg in 2002 with participants including more than a hundred heads of state and government and tens of thousands of government representatives and non-governmental organizations.

Around the middle of the decade, European leaders set a new target: Strategy 20–20-20. Launched in 2005 (but adopted in 2008), the aim of the Strategy was to ensure that by 2020, the use of fossil fuels would be reduced by 20% while the share of renewable energies would increase with 20%. Consequently, the decrease of greenhouse gas emissions was estimated also by 20%.

Architecture and Environment

The ideas of eco-friendly approaches diffused in the building industry as well. Green buildings are new buildings that comply to different assessment methods; "green" is defined by the points they score and by the certificate that is obtained, accordingly. Not all buildings that are designed in respect with the environment are defined as "green." More iconic buildings were erected in this decade. Eco-architecture, mimetic architecture, and parametric architecture are directions in architecture where technology, innovation, and imagination blend and result in astonishing results that can (if the occupants agree to use the building according to the "manual") have a lower impact on the consumption of resources or, at least, on the costs of maintenance and use.

The skyscraper at 30 St. Mary Axe in London, also—or better—known as "The Gherkin" (Fig. 2.8) owes its aerodynamical shape to the parametric design; the rounded shaft gives a better response to wind impact than a rectangular structure. Due to the huge interior shafts and the computer programmed windows, air conditioning can be reduced, if the tenants opt for it.[25]

---

[24] According to Wikipedia, Al Gore is *"an American politician, businessman, and environmentalist"*.

[25] Which they didn't, at least by 2015: "Although the Gherkin can supply an outstanding energy saving plan for all tenants, most of tenants do not choose it. "We try to sell the whole energy package to every new tenant through their M&E consultant," says Steve Brown, the building services manager. Unfortunately his exhortations have so far fallen on deaf ears, as all tenants except for Swiss Re have opted for full, all-year-round air-conditioning" [26].

Fig. 2.8 The Gherkin—30 St. Mary Axe, London. Architecture Foster and Partners. (Photo Richard Jones, CC BY-SA 2.0 DEED. Source: https://www.flickr.com/photos/richardjo53/7832245742)

The first major eco-friendly low-energy-emission concept residential development was built in London: BedZED,[26] the Beddington Zero Energy Development (Fig. 2.9). It began as a zero-carbon eco-village, and it resulted as a community with 100 homes, office space, a college, and corresponding facilities.

---

[26] More on this topic in Chap. 5.

Fig. 2.9 BedZED, Beddington zero energy development. Design team: Bill Dunster Architects, Arup, BioRegional, Ellis & Moore Consulting Engineers. (Photo Tom Chance, CC BY 2.0. Source https://en.wikipedia.org/wiki/BedZED#/media/File:BedZED_2007.jpg)

One of the things emphasized by the team was that all the benefits in terms of sustainability were results and not aims; the target was to provide good-quality living with the most efficient use of resources.

The 2010s

This decade, defined by Fawkes as the "energy efficiency phase",[27] was considered a period of consolidation of the idea that *energy efficiency* is required as a means of meeting climate targets imposed by the UN. However, at least in the UK, Dr. Steven Fawkes observes that "the value of non-energy benefits such as increased sales, increased health and well-being, as well as macro-benefits such as job creation have been recognized but have only just started to be valued. […] The value and importance of non-energy benefits […] are far more strategic to organizations than just energy saving" [1].

---

[27] According to Dr. Steven Fawkes.

Political and Environmental Approaches

During this decade, in the field of sustainable development two top-level events took place: the United Nations Conference on Sustainable Development—or Rio + 20—and the Paris Agreement, in 2015.

The Conference in Rio de Janeiro, held in June 2012, benefited (as did the previous events organized by the UN) of representatives at the highest level [27]: "world leaders, along with thousands of participants from governments, the private sector, NGOs and other groups."

The organizers aimed:

– To "focus on two themes:

  (a) a green economy in the context of sustainable development poverty eradication;
  (b) the institutional framework for sustainable development."

– To produce a political document where measures for implementing sustainable development would be provided.

At the end of 2015, France hosted a United Nations climate change conference where a new treaty was adopted, replacing the Kyoto Protocol. The target of the Paris Agreement was to reduce and maintain the global temperatures "to well below 2 °C above preindustrial levels and to pursue efforts to limit the temperature increase to 1.5 °C" [28].

The site of the European Environment Agency [29] makes a synthesis of the 25 years of policies in the field of sustainability, stating that "Over the period 1990–2016, the energy efficiency of end-use sectors improved by 30% in the EU-28 countries at an annual average rate of 1.4%/year. These advancements were driven by advancements in the industry sector (1.8%/year) and the household sector (1.6%/year). However, half of the efficiency gains achieved through technological innovation in the household sector were offset by the increasing number of electrical appliances and by larger homes."

Economical Approaches

In the past decade—and continuing nowadays—the policies to enhance the use of alternative energy systems continued: from efforts to raise public awareness carried out by media to economic and environmental politics that include national programs which promote and facilitate (by tax deductions or incentives) the installation of solar systems, wind turbines, geothermal heating and cooling systems, cogeneration facilities, and (maybe) other less known systems. Therefore, emphasis was put on the research and development of these systems in producing energy that can replace the traditional sources.

In the building industry, the integration of alternative energy producing systems as building components boosted.

Beside the integration of systems that produce energy from alternative sources (sun and wind), bringing back the vegetation into the cities from where it was banished became a more frequent approach. Collaboration between great architects and horticulturists led to the development of vegetated facades in the urban landscape. The Caixa Forum in Madrid (Fig. 2.10), designed by Herzog & De Meuron, displays a living, vertical wall that is very much like a vegetal painting, created by Patrick Blanc.

Fig. 2.10 The Caixa Forum, Madrid. Architects Herzog & De Meuron, Le Mur Végétal system by Patrick Blanc. (Photo: Fernando, CC BY-SA 2.0 DEED. Source: https://www.flickr.com/photos/fernand0/2553273925)

Today and the Predictable Tomorrow (2020–2030)

Phase 6 in Dr. Steven Fawkes' analysis was defined as a phase where efficiency is the resource.[28] In 2016 he updated his "brief history of energy management" foreseeing that "It is possible that this period will be a period of energy abundance

---

[28] "Efficiency as a resource phase".

globally, with oil, gas, renewables and efficiency all being available." The reality (so far) contradicted his optimistic views: the European Union adopted, in early 2020 "the green deal" that plans to gradually withdraw the use of the fossil fuels (in favor of energy provided by renewable sources), in an effort to decrease the greenhouse gas emissions by at least 55% by 2030 (compared with 1990). Green Plans or Green Deals were proposed throughout the whole world, in the past 20 years, under several names but with the same targets: to reduce the climate change (by limiting warming to 1.5 °C), to diminish its consequences, including providing "environmental justice" for the poor communities.

According to the European Union "Collectively, buildings in the EU are responsible for 40% of our energy consumption and 36% of greenhouse gas emissions, which mainly stem from construction, usage, renovation and demolition" [30]. The percentage is very similar in the USA: "The buildings sector accounts for about 76% of electricity use and 40% of all U. S. primary energy use and associated greenhouse gas (GHG) emissions, making it essential to reduce energy consumption in buildings in order to meet national energy and environmental challenges" [31].

This statement is, however, not entirely true (or it needs some extra explanations): buildings are not responsible for anything; they represent a shelter where specific activities are carried out: living, working, entertainment, etc. Therefore, they should not be blamed. A building with no activity produces nothing; an abandoned building uses no energy by itself; a ruin produces nothing (actually, they do produce something: dust and debris). Therefore, the statement should be formulated pointing toward the ones who use the energy to produce whatever the occupants produce.

Climate Change Approaches Over the Past Century

Even if the preoccupation of scientists on climate change was rather discrete and punctual in the last 60 years of energy and environmental policies, they had an explosive/exponential growth over the past two decades.

Nevertheless, the discussion about climate warming or cooling is at least one hundred years old [32]: in November 1922 *The Arctic seems to be warming up*" according to the Monthly Weather Review.

George Nicolas Ifft continues:

"Reports from fishermen, seal-hunters and explorers who sail the seas about Spitzbergen and the eastern Arctic, all point to a radical change in climatic conditions, and hitherto un-heard-of high temperatures in that part of the earth's surface. [...] The oceanographic observations have, however, been even more interesting. Ice conditions were exceptional. In fact so little ice has never been before noted. The expedition all but established a record, sailing as far north as 81° 29′ in ice-free water. This is the farthest north ever reached with modern oceanographic apparatus. [...] In connection with Dr. Hoel's report, it is of interest to note the unusually warm summer in Arctic Norway and the observations of Capt. Martin Ingebrigtsen, who has sailed the eastern Arctic for 54 years past. He says that he first noted warmer conditions in 1918, that since that time it has steadily gotten warmer, and that today the Arctic of that region is not recognizable as the same region of 1868 to 1917.

> *Many old landmarks are so changed as to be unrecognizable. Where formerly great masses of ice were found, there are now often moraines, accumulations of earth and stones. At many points where glaciers formerly extended far into the sea, they have entirely disappeared."*

In the 1970s, scientists balanced between global cooling and global warming.

Rasool and Schneider [33] predicted an imminent ice age, based on computer modeling and the effects of aerosols and CO_2:

> *"Effects on the global temperature of large increases in carbon dioxide and aerosol densities in the atmosphere of Earth have been computed. It is found that, although the addition of carbon dioxide in the atmosphere does increase the surface temperature, the rate of temperature increase diminishes with increasing carbon dioxide in the atmosphere. For aerosols, however, the net effect of increase in density is to reduce the surface temperature of Earth. Because of the exponential dependence of the backscattering, the rate of temperature decrease is augmented with increasing aerosol content. An increase by only a factor of 4 in global aerosol background concentration may be sufficient to reduce the surface temperature by as much as 3.5°K. If sustained over a period of several years, such a temperature decrease over the whole globe is believed to be sufficient to trigger an ice age."*

Schneider later admitted he overestimated the effect of aerosols and underestimated the contribution of CO_2. Eventually he was a consultant and an advisor for several administrations and eventually received a collective Nobel Peace Prize, in 2007, for the joint efforts of four generations of contributors in the IPCC[29] Intergovernmental Panel on Climate Change [34].

Stephen Schneider defined an ethical problem of the researcher/scientist [35]:

> *"On the one hand, as scientists we are ethically bound to the scientific method, in effect promising to tell the truth, the whole truth, and nothing but—which means that we must include all the doubts, the caveats, the ifs, ands, and buts. On the other hand, we are not just scientists but human beings as well. And like most people we'd like to see the world a better place, which in this context translates into our working to reduce the risk of potentially disastrous climatic change. [...] That, of course, entails getting loads of media coverage. So we have to offer up scary scenarios, make simplified, dramatic statements, and make little mention of any doubts we might have. This 'double ethical bind' we frequently find ourselves in cannot be solved by any formula. Each of us has to decide what the right balance is between being effective and being honest. I hope that means being both."*

Scientists are supposed to have doubts and to question everything and start over again. Dr. Schneider can serve as an example in this case, as he predicted global cooling and global warming at a difference of 3 years.

Dr. Liv Grjebine defined doubt in science as "a feature, not a bug. Indeed, the paradox is that science, when properly functioning, questions accepted facts and yields both new knowledge and new questions—not certainty" [36].

Scientific doubt was tackled by René Descartes, as early as the seventeenth century in a philosophic quintessence: *"Dubito, ergo cogito; cogito, ergo sum; sum, ergo Deus est!"*

---

[29] Intergovernmental Panel on Climate Change, of the United Nations Environment Program and the World Meteorological Organization, established in 1988.

Several computer climatic models have been proposed and tested during the past few decades but they proved to be exactly what they are called: models.

They predict, with more or less accuracy, the temperatures in the future, based on calculations, not on historic values, as the temperature measurements have been registered only for around one hundred years:

"The ultimate test for a climate model is the accuracy of its predictions. But the models predicted that there would be much greater warming between 1998 and 2014 than actually happened. If the models were doing a good job, their predictions would cluster symmetrically around the actual measured temperatures. That was not the case here; a mere 2.4 percent of the predictions undershot actual temperatures and 97.6 percent overshot, according to Cato Institute climatologist Patrick Michaels, former MIT meteorologist Richard Lindzen, and Cato Institute climate researcher Chip Knappenberger. Climate models as a group have been "running hot," predicting about 2.2 times as much warming as actually occurred over 1998–2014. Of course, this doesn't mean that no warming is occurring, but, rather, that the models' forecasts were exaggerated" [37].

While not denying global warming as a natural and cyclic phenomenon, a global network of over 1900 scientists and professionals launched the World Climate Declaration *There is no climate emergency* [38]. Illustrious names like Ivar Giaever,[30] John Clauser,[31] Richard Lindzen,[32] and other specialists in environmental sciences, meteorology, or geology state that "Climate science should be less political, while climate policies should be more scientific. Scientists should openly address uncertainties and exaggerations in their predictions of global warming, while politicians should dispassionately count the real costs as well as the imagined benefits of their policy measures."

Nobel Prize winner physicist John Clauser considers that the anthropogenic activities are not the source of the changes of the climate. In 2023 he declares[33]: "In my opinion, there is no real climate crisis. There is, however, a very real problem with providing a decent standard of living to the world's large population and an associated energy crisis. The latter is being unnecessarily exacerbated by what, in my opinion, is incorrect climate science" [40, 41]. The number of skeptics that question the relation between global warming and the production of dangerous greenhouse gases (CO_2 or CH_4 due to anthropogenic activities and those of other living creatures for that matter[34][35]) seems to grow.[36]

---

[30] Ivar Giaever was awarded the Noble Prize in Physics in 1973 (shared with Leo Esaki and Brian Josephson).

[31] John Clauser was awarded the Nobel Prize in Physics in 2022.

[32] Richard Lindzen, professor at the MIT Center for Global Change Science, expert in climate science since the mid-1970s.

[33] Unfortunately—and inexplicably, in our opinion—his speech at the International Monetary Fund was cancelled [39].

[34] In [42].

[35] In [43].

[36] The Petition "There is no Climate Emergency" began with the signature of 500 scientists in September 2019, numbers over 1900 now and is still growing.

Dr. Richard Lindzen affirmed in 2023 that the idea of using a mean value as term of reference (maximum 1.5 °C) for the temperature increase at the level of the whole planet is not operable, as there are "dozens of climate regimes"; he also observes that this increase of 1 °C (since the invention of the steam engine which meant harvesting fossil fuels) is accompanied by the greatest welfare of the humans. People are led to believe that "by getting rid of CO_2 they're doing something virtuous. You know, as I occasionally pointed out, let's say somebody came up with a good device and they could get rid of about 60–70% of the CO2 in the atmosphere; what would be the result? The result would be we'd all be dead. That's a very peculiar pollutant, one that we can't live without" [44].

On the other hand, the opinions of scientific personalities who consider that these are the last days of humanity are also strong: Professor Guy McPherson, a leading global voice on abrupt climate change, believes humans will be extinct by 2026 and considers that the reports of the Intergovernmental Panel on Climate Change are too conservative, as "Every single one of them so far, each of the six, has concluded the previous assessment was too conservative. Now even the IPCC admits we are in the midst of abrupt and irreversible climate change. Irreversible" [45].

What seems to be undoubtedly is the fact that CO_2 increased, inexplicably, in pre-industrial periods as well, when major anthropogenic activities did not occur: during the last glacial period, between 23,000 BC and 8000 BC the level of CO_2 did not follow the line of the temperature lag/growth [46].

In less than three decades, the debate on climate change and who (or what) is causing it amplified (and became confusing): while until the past decade the fact that the water vapors represented by far most of the greenhouse gases (and the statement is—or was, at the time—supported by outstanding names in the fields of meteorology, climatology, biology, physics), the optics has changed in the last few years and, oddly enough, some sites (and researches) present today different (nuanced) components and ratios: firstly, a difference is made between water vapor and clouds (which are condensed water vapors) and secondly the percentages of CO_2 declared in the early 2000s differ substantially from those declared today.

The late (and regretted) Dr. Tim Ball wrote, in 2016:

> "Water vapour is the most important greenhouse gas. This is part of the difficulty with the public and the media in understanding that 95% of greenhouse gases are water vapour. The public understand it, in that if you get a fall evening or spring evening and the sky is clear the heat will escape and the temperature will drop and you get frost. If there is a cloud cover, the heat is trapped by water vapour as a greenhouse gas and the temperature stays quite warm. If you go to In Salah in southern Algeria, they recorded at one point a daytime or noon high of 52°C—by midnight that night it was − 3.6°C. That's a 56-degree drop in temperature in about 12 hours. That was caused because there is no, or very little, water vapour in the atmosphere and it is a demonstration of water vapour as the most important greenhouse gas" [47].

John Reilly[37] wrote the following statement in an email [20]: "Water vapor is the most important greenhouse gas and natural levels of [carbon dioxide, methane and nitrous oxide] are also crucial to creating a habitable planet."

---

[37] John Reilly, professor at MIT.

Feineman considers that "it is fairly accurate to say that water vapor supplies close to half of the total greenhouse effect, clouds and carbon dioxide each a little under a quarter, and all others just under a tenth" [48] and continues with the following (fascinating) hypothesis: "Some research has looked at what would happen if carbon dioxide were removed from the atmosphere. Loss of the carbon dioxide cools the planet, but that condenses some of the water vapor, which cools the planet more, and the Earth turns into an ice-covered snowball." Fascinating, because the problem is tackled from a different angle: if CO_2 is removed from the atmosphere, life on planet Earth will be extinct.

After all the pros and cons and the ongoing debate (that has a political and social component, not only a scientific stake), there are some things that should be clear, about greenhouse gases:

– The greenhouse gases are essential for life on planet Earth: had it not been for the greenhouse gases, the temperature on Earth would have dropped to as low as $-18\ °C\ (-0.4\ °F)$.[38]
– 95% of the greenhouse gases are water vapors[39]; the remaining 5% are represented mainly by natural occurring gases—carbon dioxide (CO_2), methane (CH4), nitrous oxide (N2O), and ozone (O3)—and of course fluorinated gases produced by the anthropogenic activities: hydrofluorocarbons, perfluorocarbons, sulfur hexafluoride, and nitrogen trifluoride.
– CO_2 should not be regarded as a pollutant; without it life on the planet Earth would stop: plants require CO_2, sunlight, and water and in the process of photosynthesis produce (in fact, eliminate) oxygen. CO_2 is essential for life on the planet as it represents "the food of plants"—as defined beautifully by Dr. John Christy.[40]
– There were other periods of global warming throughout history[41] and they were not related to anthropogenic activities and to CO_2: according to Jan Esper's studies on tree rings, during the Roman period the temperature might have been with 0.6 °C higher than what we have today [50].

However, humans are responsible for massive pollution and deforestation, even if it is done in the name of saving the planet: according to [51] "One of the only instances where cutting down trees can be environmentally friendly is if trees were being removed to maximize solar panel exposure. Removing trees to get more exposure to solar panels is actually very environmentally friendly." REALLY? Of course,

---

[38] In [49].

[39] See Footnote 16.

[40] John Christy, climate scientist and professor of atmospheric science at the University of Alabama in Huntsville.

[41] In fact there is a periodicity of the warm periods and cold periods, not at a geological scale but at the scale of history; it is recognized that there was a Medieval Warm Period (1100–1300 AD also known as the Little Climatic Optimum), followed by the Little Ice Age that ended at around 1900 A.D. Therefore, the panic of a new cold period was not surprising in the first half of the twentieth century.

calculations can prove almost anything in specific hypothesis (like this one that compares trees with equipment) but cutting trees for installing PV panels or wind farms is not eco-friendly, no matter what theory supports it.

For that matter, the praised furniture factory The Plus by Vestre (Fig. 2.11) claims that it "is well-placed to become the world's first project of its type to achieve the highest BREEAM environmental classification: "Outstanding". The factory has been built as a 'plus house', which means that the building's energy consumption is 60 percent lower than equivalent conventional factories, while releasing 55 percent less greenhouse gases than similar buildings. If you include the manufacturing equipment in the equation, energy consumption in The Plus is reduced by 90 percent" [52].

Fig. 2.11 The Plus by Vestre Architects: Bjarke Ingels Group (BIG). (Photo: Google Earth)

In order to make room for the factory, the infrastructure and everything else that is needed in such an industrial building, trees were massively cut. Placing photovoltaic panels does not compensate for the cut trees as there are two altogether different things that trees and PV panels do: one encapsulates/feeds on CO_2, produces oxygen, absorb electromagnetic radiation while PV panels produce electricity (which reached an efficiency of 20% in practice and which still raise questions about the possibilities of recycling but these aspects are rarely tackled).

This is, in our opinion, stretching the boundaries of "eco-friendly approaches" that aim to draw attention upon and increase the use of these alternative energy systems.

In the same context the announcement published by The Telegraph in 2023 is mind blowing: "Almost 16 million trees have been chopped down on publicly owned land in Scotland to make way for wind farms" [53].

The anthropogenic approach (of cutting down forests and trees) is probably devastating for the planet and should be dealt with, considering all the environmental consequences (from changing of the microclimate and destroying the flora and fauna of the area to the role of CO_2).

And these issues need to be addressed firmly with effective/efficient measures, as their impact on the planet is destructive.

Environmental Policies in Buildings and Building Products

For nearly two decades, the European Union launched and systematically updated Directives that aim to keep under control the use of energy in buildings. In December 2021, the Commission proposed a revision of the Energy Performance of Buildings Directive that sets new and ambitious targets regarding the buildings—either new or old—that match the provisions and expectation of the European Green Deal. "As of 2021, all new buildings must be nearly zero-energy buildings (NZEB) and since 2019, all new public buildings should be NZEB. When a building is sold or rented, energy performance certificates must be issued and inspection schemes for heating and air conditioning systems must be established" is stated on the Energy page of the site of the European Commission.

The term nearly zero-energy buildings was defined and launched in the 2010 edition of the Directive on the energy performance of buildings [54] as follows: "a 'nearly zero-energy building' means a building that has a very high energy performance, as determined in accordance with Annex I. The nearly zero or very low amount of energy required should be covered to a very significant extent by energy from renewable sources, including energy from renewable sources produced on-site or nearby." Annex I presents the main characteristics that need to be taken into consideration in establishing a methodology that defines and develops these building typologies: "The energy performance of a building shall be determined on the basis of the calculated or actual annual energy that is consumed in order to meet the different needs associated with its typical use and shall reflect the heating energy needs and cooling energy needs (energy needed to avoid overheating) to maintain the envisaged temperature conditions of the building, and domestic hot water needs."

"A very high energy performance" is, however, a diffuse qualitative measure; very high compared to what? It is understandable that performance is related to the economic development and local resources of each country and therefore things cannot be considered in a Procrustean manner. Therefore, "for new buildings,

Member States shall ensure that, before construction starts, the technical, environmental and economic feasibility of high-efficiency alternative systems such as those
listed below, if available, is considered and taken into account:

(a) decentralised energy supply systems based on energy from renewable sources;
(b) cogeneration;
(c) district or block heating or cooling, particularly where it is based entirely or
     partially on energy from renewable sources;
(d) heat pumps."

The Directive stipulated that "The energy performance of a building shall be determined on the basis of the calculated or actual annual energy that is consumed in
order to meet the different needs associated with its typical use" and therefore methodologies were established, to *calculate* the energy performance of the buildings
rather than to *verify* the annual consumption (which would give the real stat of the
art). In this sense, Annex I of the abovementioned document indicated a minimum
of "aspects": design features that have to do with the positioning and orientation of
the building, characteristics of the building materials (thermal capacity and insulation) and overall configuration (with emphasis on the thermal bridges), passive
design measures (passive heating and cooling, natural light and ventilation, solar
systems and protection), as well as installations (heating and hot water supply, air
conditioning, lighting). Some of them provide, in our opinion, only qualitative values ("this is better than that"), and beyond calculations, empirical validations are
mandatory but this research step (essential we would say), apart from assigning
financial resources, would also be time consuming and the artisans of the energy
efficiency Directive have deadlines to meet and a time-table to stick to. The 2018
amending Directive [55] as well as other proposals, like the REPowerEU, presented
in 2022 [56] set more and more ambitious targets of energy efficiency by saving or
reducing energy consumption.

   However, the most recent version of the Directive (not adopted yet but under
discussions) has taken into consideration the evaluation of the building from an
interdisciplinary perspective; while the previous editions focused exclusively on the
energy performance of the building, this version evaluates other requirements that
the building should fulfill: fire safety, accessibility, air quality, and so on. Passive
solutions are mentioned: vegetated roofs and facades, the use of solar passive gain
in different assemblies that diminish the quantity of energy that needs to be used,
hence decreasing the overall consumption of energy. Indirectly, architectural design
is emphasized, as all the passive means are in the "yard" of the architect. And, in our
opinion, this is a correct approach: the architect is the creator of a territorial or of an
urban "sculpture," in which humans live and work, the orchestrator that gives a
theme to each of the specialists involved in the act of creation: other architects,
engineers, environmentalists, geologists, botanists, etc.

   Together—and only together—the solution is shaped and polished until it is fit
for construction. In this endeavor the passive and active means blend into the final

result; each has its own importance and determines a specific reaction of the building.

Omitting or emphasizing only some aspects would lead to an unbalanced response of the construction, as Jan Rosenow and Ray Galvin demonstrate in [57]: "a substantial body of literature covering several decades of energy research suggests estimated savings in evaluations are often higher than actual, measured savings." Furthermore, as Minna Sunikka-Blank and Ray Galvin conclude in [58] "The German datasets discussed above indicate that the real measured household heating energy consumption could be on average 30% lower than calculated."

According to the Directive, specific methodologies should be established to estimate the energy consumption of a building according to its components and equipment.

So far, these methodologies have not been validated by real-life measurements and therefore the results are not always real.

Contemporary Policies Regarding Building Components and Materials

During the last decade, the building sector became the area of study of different associations that evaluate the impact of building materials on the environment—in relation, obviously, with the carbon emissions.

Launched in December 2015, the Global Alliance for Buildings and Construction[42] presented its first Global Status Report for Buildings and Construction 2016: Towards Zero-emission, Efficient and Resilient Buildings. It was elaborated by the International Energy Agency under the coordination of the United Nations. The site of the organization [60] hosts all the submitted reports, one for every year.

Comparing the similar data of the reports (and not taking into consideration that adding the appropriate shares, the result exceeds 100%), Table 2.1 shows that buildings—residential and non-residential—account yearly for 30% of the final energy consumption. Construction industry consumes between 5 and 12% (an explanation can be that from 1 year to another some industrial processes were located to "other industry").

---

[42] According to [59]"GlobalABC functions as an umbrella or meta-platform—a network of networks—that brings together initiatives and actors focusing on the buildings and construction sector."

Table 2.1 Share of global final energy consumption by sector (based on the data provided by the global alliance for buildings and construction

year	Buildings			Construction industry	Transport	Other industry	Other
	Residential	Nonresidential	Total				
2016 (data 2014)	NA	NA	31	9[a]	27	28	6
2017 (data 2015)	22	8	30	6	28	31	5
2018 (data 2017)	22	8	30	6	28	32	4
2019 (data 2018)	22	8	30	6	28	32	4
2020 (data 2019)	22	8	30	5	28	32	5
2021 (data 2020)	22	8	30	6 + 6	26	26	6
2022 (data 2021)	21	9	30	4 + 1 + 3	26	31	6

[a]Related to iron, steel, and cement industry

While the solar technologies were promoted and, despite their high production prices, the energy and environmental policies had an important contribution in their implementation as building components (building integrated photovoltaics BIPV, building integrated solar thermal BIST), the traditional building materials were evaluated from a new perspective: the energy that is used to produce them.

Steel, iron, cement, brick, and glass are common (and essential) building materials but, in terms of energy consumption they are all energy intensive, as they require high temperatures (above 1000 °C) in the process of production. They are all also important anthropogenic sources for carbon emission and other polluting agents:

– The fabrication of cement accounts for about 5% of the yearly global anthropic carbon emissions [61].
– The aluminum industry generates around 2% of the yearly anthropic carbon emissions [62].
– Brick kilns release each year CO_2 emissions that represent 2.7% of the total anthropogenic emissions [63].
– Brick firing consumes around 24 million tons of coal/year [64].
– Glass is accounted for producing 95 million tons CO_2 worldwide each year [65], representing around 0.27% of the total CO_2 anthropogenic emissions.

By adding "other building and construction industry" [66] results that the most commonly used building materials contribute to the total CO_2 *anthropogenic emissions* with around 15% (Table 2.2).

Table 2.2 Share of global energy-related anthropogenic CO_2 emissions by sector (based on the data provided by the Global Alliance for Buildings and Construction)

year	Buildings					Construction industry	Transport	Other Industry	Other
	Residential %		Nonresidential %						
	Direct	Indirect	Direct	Indirect	Total				
2016 (data 2014)					–	–	–	–	–
2017 (data 2015)	6	11			28	11	22	30	9
2018 (data 2017)	6	11	3	8	28	11	23	32	6
2019 (data 2018)	6	11	3	8	28	11	23	31	7
2020 (data 2019)	6	11	3	8	28	10	23	32	7
2021 (data 2020)	6	11	3	7		$(10 + 10)^a$	23	23	6
2022 (data 2021)	6	11	3	8		$(6 + 3 + 6)^b$	22	30	8

[a] Building construction industry + other construction industry [66]
[b] Concrete, aluminum, and steel + bricks and glass + other building and construction industry [66]

So what should be done? Circular economy, wherever applicable, can be one answer. And a good one too!

One of the measures of decreasing useless consumption (and matching the criteria of diminishing the carbon footprint) can be the recycling of products. It is not a new concept either; in fact we can call it a recycled concept: in Romania,[43] for instance, in the communist regime, "the three R-s" were in force: repair, recycle, and reuse. Other countries had politics on recycling products (i.e., bottle bills, through which every empty can or bottle can be reimbursed) as early as 1970s–1980s. Today, among the actions embraced by the climate action activists, the number of "R"s has increased from 3 to 5: *refuse, reduce, reuse, repair, and recycle* (in this order). However, the principle of the Rs—no matter if 3, 5, or 7—is in fact imported from the circular economy that is based on the idea that there is (or should be) no waste: what is a by-product deriving from an activity should become a feedstock for another activity. Circular economy—the "no waste" economy—grew as an alternative to the 1955s consumerist "throwaway living" trend that promoted single-use

---

[43] The author of the book is Romanian.

items as a norm [67]. Considering that consumerist trend began after World War 1, it may be a reaction of psychological defense after the fear, poverty, frustration, privation, and scarcity represented by the war. The "crazy years" came as a liberation, a "carpe diem" attitude: "live the moment." The throwaway attitude of the 1950s might have the same cause—trauma following World War 2—but with a different expression.

Currently its principles can be defined by seven keywords: redesign/rethink, reduce, reuse, repair, renovate/refurbish, recycle, and recover[44] [68, 69].

As strange as it may seem, preoccupations for a "cradle to cradle" approach of the management of the resources is probably almost as old as mankind: in the world of the traditional should be the local building tradition everything was recovered and used with the maximum efficiency. Not to save the planet but to save money. Indirectly the users of the buildings made the most of the available resources with the minimum costs, while providing a better living standard. The energy that was consumed was in fact the difference between the one provided naturally, by the configuration of the building and spaces (through passive design) and the one required for providing the indoor conditions of temperature, humidity, and light specific to the historic period in discussion.

While throughout history people repaired, reconsidered, and used to the maximum their products, organized measures of recycling have been documented as early as the eleventh century in Japan where paper was recycled [68, 70]. Other materials that were recycled in the past include metal, textiles, and ceramics. Only the consumerist society of the second half of the twentieth century, aiming to stimulate the production of goods, changed the mentalities from product for a lifetime (a car, a washing machine, a refrigerator that would last for at least 20 years) to products that would be superseded by newer models in few years (merely months, if discussing of smartphones). The philosophy applies to buildings as well: the installations have a lifetime expectation of 10 years; products for finishing replace the massive bricks, wood, or stone with composite panels that are lighter, thinner, and easy to fabricate, to install, or to replace. Therefore the idea of consumerism slides to the building industry as well.

Since Socrates, with the famous paradox "The only true wisdom is in knowing you know nothing," humanity (must have) learned that doubt and interest are the engines to new experiences, discoveries, and better understanding. What we learn about the climate of our planet and how it influences us or how we influence it is in many aspects unknown and the models (that are based on current information and several hypotheses) are questioned by specialists, as they are totally different from our estimations[45] and do not match with the real-life measurements or with the historic measurements (that are not too accurate either).

---

[44] More on this topic in Chap. 3.

[45] "The sun is more surprising than we knew [...] We thought we had this star figured out, but that's not the case" said Mehr Un Nisa in August 2023 [71].

Beyond all the charts and papers that draw alarm signals of how we warm the planet, the reality of pollution and wasting our resources—which nobody is doubting—is evident as:

1. The building materials that are employed in the construction process have been obtained with a large energy consumption (and with anthropogenic fluorinated gas emissions) and no matter what changes or improvements will be researched, the process of production of the most common building materials will be carried out at very high temperatures, hence with high energy consumption (although there have been some experiments of alternative technologies in the field of metallurgy).
2. The building should not be considered responsible of anything; it is our responsibility, as specialists, to design it in a manner that would optimize the energy consumption. Like in the case of any product—and the building is a product—it should perform well with less energy and—eventually—it might need a manual of use. An important design parameter should be the local building tradition and know-how, as it seems that people survived without, for instance, the HVAC systems for millenia.[46]

Summarizing a century in a paragraph, some things seem to be certain: buildings should be designed taking into consideration that to work (properly), energy needs to be spent efficiently. In other words, one of the requirements that a building should fulfill is how much it consumes. It was never stated using quite these words in the past but the building, from its geometry and space to the last component, was "tuned" in order to achieve this specific, unstated requirement: *how much will it cost in use*. And the question did not address the rich few but to most of the population (who permanently watches the expense balance). This is why the same principles apply in the traditional housing all over the world; or why solutions that were used 5–6.000 years ago were "reinvented" recently: they all have the same denominator: they use the force and the rules of nature as a tool of design, minimizing the consumption of energy resources while providing the parameters of comfort. In other words, designing with nature and not defying it (although such statement-buildings or cities exist and will always be built, as a message of power and wealth).

It was a principle long before any of the contemporary methodologies were in force. In fact, it is no need to draw a methodology for something as logical as the fact that everyone wants to pay less for a better living standard.

The building is the product with the longest life span. There are millennial buildings that not only stand but are still functioning. While not denying the political will and the strategies, it must be emphasized that energy-efficient buildings were created long before the need to assess their energy performance from the point of view of the energy that they use. "The will" that Ludwig Mies van der Rohe was speaking about is, in our times, based on "the quest" of intertwining the peoples' needs with

---

[46] More on this topic in Chap. 4.

the planets' capacity to achieve them without unbalancing either of the two. The means of accomplishing these needs—the "engine"—is energy. And this epoch has the technological means for producing energy from alternative resources that can be embedded in building products. Hence, why not use them?! As means not as targets, as the buildings should continue to fulfill their primordial role: a safe (and cozy) shelter to the occupants.

A good start would be to understand the traditional architecture; everywhere in the world the same principles can be found and they are the fundament of most of the architectural approaches of the past 50 years: bioclimatic, eco-architecture, and even solar architecture to some extent are rooted in the building concepts of the vernacular. And, going further, biomimetic architecture has its roots in the "humble" world of other creatures.

References

1. https://www.onlyelevenpercent.com/a-brief-history-of-energy-efficiency. Accessed October 2022 and August 2023.
2. United Nations Environmental Programme, Environmental Law Guideline Programme, Stockholm Declaration, p. 6. https://wedocs.unep.org/bitstream/handle/20.500.11822/29567/ELGP1StockD.pdf, Accessed 2023.
3. Dabija, A.-M. (2021). *Alternative envelope components for energy-efficient buildings*. Springer Nature. https://link.springer.com/book/10.1007/978-3-030-70960-0
4. https://earthrivergeothermal.com/history-of-geothermal-systems/
5. Janssen, J. E. (1999, September). The history of ventilation and temperature control. *ASHRAE Journal*, 47–52.
6. https://timeline.com/sick-building-syndrome-cc82f76a07e9 accessed August 2023
7. Crawford, J. O., & Bolas, S. M. (1996). Sick building syndrome, work factors and occupational stress. *Scandinavian Journal of Work, Environment & Health, 22*, 243–250. https://doi.org/10.5271/sjweh.138
8. https://www.cdc.gov/about/pdf/facts/cdcdiscovery/discoveries-series%2D%2D-legionnaires.pdf
9. https://www.cdc.gov/vitalsigns/legionnaires/
10. https://thelegionnaireslawyer.com/history-legionnaires-disease/
11. Muraca, P., Stout, J. E., & Yu, V. L. (1987, February). Comparative assessment of chlorine, heat, ozone, and UV light for killing legionella pneumophila within a model plumbing system. *Applied and Environmental Microbiology*, 447–453.
12. https://www.osha.gov/legionnaires-disease/control-prevention
13. Archibald, J., Building integrated solar thermal roofing systems history, current status, and future promise, Proceedings of the solar conference, 1999—bestroofersinmyarea.com
14. http://history.alberta.ca/energyheritage/energy/solar-power/modern-photovoltaic-power.aspx
15. https://www1.eere.energy.gov/solar/pdfs/solar_timeline.pdf
16. https://www.smithsonianmag.com/sponsored/brief-history-solar-panels-180972006/
17. https://www.udel.edu/udaily/2018/april/in-memoriam-karl-w-boer/
18. https://www.nasa.gov/centers/armstrong/news/FactSheets/FS-054-DFRC.html
19. https://www.nma.gov.au/explore/collection/highlights/quiet-achiever-solar-car
20. https://globalchange.mit.edu/news-media/in-the-news/greenhouse-gases-water-vapor-and-you
21. https://unfccc.int/kyoto_protocol

22. https://worldgbc.org/about-us/our-mission/
23. Stephens, S. (2000–03). Fox & Fow e creates a collage in FOUR TIMES SQUARE, using skyscrapers past and present and a touch of "green", Architectural Record, p. 95. https://usmodernist.org/AR/AR-2000-03.pdf
24. https://www.nytimes.com/1992/03/15/nyregion/solar-energy-a-hope-of-the-70s-and-80s-suffers-decline.html
25. https://www.ambasz.com/about
26. https://lianaxiong.weebly.com/blog/april-10th-2015
27. http://www.uncsd2012.org/about.html
28. https://www.ipcc.ch/sr15/faq/faq-chapter-1/
29. https://www.eea.europa.eu/data-and-maps/indicators/progress-on-energy-efficiency-in-europe-3/assessment
30. https://commission.europa.eu/news/focus-energy-efficiency-buildings-2020-02-17_en
31. Quadrennial Technology Review An Assessment Of Energy Technologies And Research Opportunities Chapter 5: Increasing efficiency of building systems and technologies September 2015, in www.energy.gov/sites/prod/files/2017/03/f34/qtr-2015-chapter5.pdf
32. https://journals.ametsoc.org/view/journals/mwre/50/11/1520-0493_1922_50_589a_tca_2_0_co_2.xml
33. Rasool, S. I., & Schneider, S. H. (1971). Atmospheric carbon dioxide and aerosols: Effects of large increases on global climate. *Science, 173*(3992), 138–141. https://doi.org/10.1126/science.173.3992.138
34. https://stephenschneider.stanford.edu/References/Biography.html
35. Schneider, S. H. (1996, August/September). Don't bet all environmental changes will be beneficial, American Physical Society. *APS News, 5*(8). https://www.aps.org/publications/apsnews/199608/environmental.cfm
36. Grjebine, L. (2020). Why doubt is essential to science. https://www.scientificamerican.com/author/liv-grjebine/
37. Henderson, D. R., & Hooper, C. L. (2017, April 4). Flawed climate models the relationship between CO2 and temperature is more complicated than the polemics suggest in hoover institution. https://www.hoover.org/research/flawed-climate-models
38. https://clintel.org/wp-content/uploads/2024/04/WCD-240411.pdf
39. https://co2coalition.org/news/cancellations-start-for-john-clauser-after-nobel-physics-laureate-speaks-out-about-corruption-of-climate-science/
40. https://www.newsweek.com/nobel-prize-winner-who-doesnt-believe-climate-crisis-has-speech-canceled-1815020
41. https://brusselssignal.eu/2023/07/nobel-prize-winning-physicists-speech-at-imf-blocked-after-questioning-climate-crisis/
42. https://www.nationalgeographic.com/animals/article/150803-cows-burp-methane-climate-science
43. https://www.reuters.com/breakingviews/war-cow-farts-is-stinky-necessary-job-2023-03-24/
44. Editor BizNews. (2023, April 21). Dr Richard Lindzen exposes the climate change movement as a fabricated, politicised power play motivated by malice and profit. https://www.biznews.com/energy/2023/04/21/climate-change-5
45. Editor BizNews. (2023, January 20). Humans will be extinct by 2026"—doom-and-gloom prophet Professor Guy McPherson on abrupt climate change. https://www.biznews.com/energy/2023/01/20/abrupt-climate-change
46. https://www.carbonbrief.org/explainer-how-the-rise-and-fall-of-co2-levels-influenced-the-ice-ages/
47. https://archive.ph/NPQ32#selection-617.1-617.856
48. Feineman, M. et al., *Courseware module of the repository of open and affordable materials*. The Pennsylvania State University, in https://www.e-education.psu.edu/earth104/node/1262
49. https://education.nationalgeographic.org/resource/greenhouse-effect-our-planet/

50. Fred Pearce, Tree rings suggest Roman world was warmer than thought https://www.newscientist.com/article/dn22040-tree-rings-suggest-roman-world-was-warmer-than-thought/
51. https://solarisrenewables.com/blog/consider-tree-removal-solar/
52. https://vestre.com/news/we-are-now-officially-opening-the-plus-the-worlds-most-environmentally-friendly-furniture-factory
53. https://www.telegraph.co.uk/politics/2023/07/19/snp-chopped-down-16m-trees-develop-wind-farms-scotland/
54. Official Journal of the European Union, vol. 53, ISSN 1725-2555, https://doi.org/10.3000/17252555.L_2010.153.eng.
55. Official Journal of the European Union, volume 61, 21 December 2018, ISSN 1977-0677, in https://eur-lex.europa.eu/legal-content/EN/TXT/?uri=OJ:L:2018:328:TOC
56. https://commission.europa.eu/strategy-and-policy/priorities-2019-2024/european-green-deal/repowereu-affordable-secure-and-sustainable-energy-europe_en
57. Jan Rosenow, J., & Galvin, R. (2013, July) Evaluating the evaluations: Evidence from energy efficiency programmes in Germany and the UK, energy and buildings, volume 62, pp. 450–458https://doi.org/10.1016/j.enbuild.2013.03.021.
58. Sunikka-Blank, M., & Galvin, R. (2012). Introducing the prebound effect: The gap between performance and actual energy consumption. *Building Research & Information, 40*(3), 260–273. https://doi.org/10.1080/09613218.2012.690952
59. https://globalabc.org/about/history-timeline
60. https://globalabc.org/our-work/tracking-progress-global-status-report
61. Worrell, E., Price, L., Martin, N., Hendriks, C., & Ozawa, M. L. (2001. Worrell). Carbon dioxide emissions from the global cement industry. *Annual Review of Energy and the Environment, 26*, 303–329.
62. https://www.weforum.org/agenda/2020/11/the-aluminium-industry-s-carbon-footprint-is-higher-than-most-consumers-expect-heres-what-we-must-do-next/
63. https://hablakilns.com/environment/environmental-impact/
64. Sameer, M. Indian Institute of Technology Bombay, Department of Energy Science & Engineering; (2003). Energy utilisation in brick kilns. Ph.D Thesis, in Mary Lissy P.N., Peter, C., Mohan, K., Greens, S., George S. In *Energy efficient production of clay bricks using industrial waste,* Heliyon. 2018 Oct; 4(10): e00891, published online 2018 Nov 2. https://doi.org/10.1016/j.heliyon.2018.e00891.
65. https://cinea.ec.europa.eu/news-events/news/how-life-reducing-emissions-glass-production-2022-03-16_en
66. United Nations Environment Programme. (2022). 2022 global status report for buildings and construction: Towards a zero-emission, efficient and resilient buildings and construction sector.
67. https://hintonswaste.co.uk/news/history-of-recycling-timeline/
68. https://urnabios.com/a-story-of-circular-design-ethics-and-sustainability-in-the-bios-urn-design/
69. https://medium.com/circular-disruption/7-circular-economy-principles-to-rethink-the-value-of-waste-9ba9a219b218
70. https://quinterecycling.org/the-history-of-waste-and-recycling/
71. https://scitechdaily.com/solar-surprise-scientists-discover-unprecedented-high-energy-light-coming-from-the-sun/

Chapter 3
From Symbolic and Sacred to Economic and Efficient: Circular Economy and the Buildings

Preamble

In all the human settlements throughout history, the buildings that reached their functional or physical end of life were—naturally—dismantled and the components that could be recuperated began a new life circuit: stone, wood, and bricks were carefully taken off decayed buildings and used in different structures: in the vicinity of ancient sites, the local population, sometimes centuries later, took the opportunity of finding dressed stone or processed wood, a stone's throw away. Today we would fit these actions into elaborated theories of green and eco-friendly approaches that aim to decrease the carbon footprint of the building products through initial embodied energy as well as recurrent/operational embodied energy. The simple question is why waste time and energy (human and horse-power) to find a quarry, cut the stone, transport it, shape it, and use it when the source of building material—the ruin—was right there, available for use?

It is what happened with many ancient sites that probably disappeared altogether but their traces can still be seen in bits and pieces, fragments that can be found in newer (but still, very old) buildings. Capitals, parts of columns, or stone stairs are integrated in walls and, in their specific way, still tell the story of their life and transformation.

© The Author(s), under exclusive license to Springer Nature Switzerland AG 2024   47
A.-M. Dabija, *Architectural Design Strategies for Saving Energy in Buildings*,
https://doi.org/10.1007/978-3-031-73541-7_3

Fig. 3.1 Ancient Orthodox church of Saint Eleftherios in Athens—capitals and stairs integrated in the walls. (Photo: Ana-Maria Dabija)

Combinations of bricks and several types, shapes, and sizes of stone support the theory that the builders used what was closest and easiest to transport and install. Athens, a flourishing city of the Antiquity, was, with its ruined buildings, the perfect and most accessible quarry for the Athenian constructors of the early Middle Ages. The church of Saint Eleftherios was literally built over the ruins of the ancient temple of Eileithyia at the turn of the thirteenth century (Figs. 3.1 and 3.2).

Fig. 3.2 Ancient Orthodox church of Saint Eleftherios in Athens—composite wall. (Photo: Ana-Maria Dabija)

The bottom line is that todays' theories on sustainability are in fact common sense approaches adopted by builders in order to simplify their activity, not complicate it. The novelty consists of the fact that these common sense and historic approaches are now put in writing and represent the core of several derived theories from sustainability to circular economy.

Theories and codes for buildings have not been invented today, by us. Building codes existed since Hammurabi[1] and the oldest surviving treatise on architecture is Vitruvius's *De architectura.*[2]

There is a probability that these were not the only books on architecture and, maybe, manuals on building technologies existed as well. If not, the oral transmission from one generation to another managed to preserve and pass on the knowledge to our times.

Symbolic Relocations of Building Components: The Grand Shrine in Ise

Buildings are more than simple shelters and should not be reduced to the materiality of their structure. Some of them are custodians of history expressed in traditions, culture, and technology. They define space. They define us. They define society. In other words, they have a material and an immaterial dimension.

Such a complex case is the Grand Shrine in Ise, Japan: with a history of more than 1300 years, with a huge symbolic and religious importance, the Grand Shrine in Ise is also an extraordinary lesson on building methods and technologies preserved and transmitted from one generation to another for over one millennia. Besides the sacred component of the assembly, the physical, secular dimension of the buildings promote, through the concept of dismantling and reusing existing components, prefabrication and standardization. In fact, Japan is probably the cradle of modulation in buildings: the surface of a room is expressed by the number of tatami mats[3] that cover it, not by length and width.

Religiously, Ise Jingu is the soul of Japan; it is a monastic assembly that counts 125 jinja[4] located on an area that, roughly, equals the center of Paris [1].

---

[1] The Code of Hammurabi is one of the world's oldest code of laws—a total of 282—written around 1754 BC. Six laws—fom 228 to 233—refer to the responsability for the consequences of the failures in buildings.

[2] Marcus Vitruvius Pollio, De architectura. Libri decem/Ten Books on Architecture, 30–20 B.C.

[3] A tatami is a traditional Japanese mat of (aproximatively) 90 x 180 cm, used to cover the floors of the rooms.

[4] The Shinto places of worship.

The (hi)story of the Grand Shrine in Ise is fascinating and worth sharing, both for its symbolic significance and for the method of passing the technological know-how through the centuries and the respectful approach toward the environment.

It is dedicated to the most revered deity of Japan, Amaterasu-Omikami. The legend says that she is a kami[5] and descends from the divine couple who gave birth to the Japanese islands and to most of the other deities.[6] Amaterasu sent her grandson to the terrestrial world, to take care of Japan. In this endeavor, he was granted the Three Sacred Treasures—a mirror, a jewel, and a sword—and was assisted by several kami. The mirror was to be placed in the Imperial palace and worshiped as if Amaterasu was there. She also gave him an ear of rice, so that the people would always have food. Ninigi-no-mikoto (the nephew) landed on Kyushu island and built his palace there. The establishment of the Japanese nation began with him and the Three Sacred Treasures were inherited, from one generation to another, by the Imperial Household. After protecting the nation from the imperial palace for a few generations, Amaterasu decided to move her symbol—the sacred mirror—in a more appropriate place that was found, through a revelation, on the banks of the Isuzugawa River (some 2000 years ago).

About 1500 years ago Toyo'uke-no-Omikami, the kami that provides food, shelter, and clothes, joined Amaterasu as her companion and food provider [2, 3], in the spaces of a shrine (the outer shrine) that was built especially for her.

Originally, shrines were temporary buildings, erected to host the deities during religious festivals. Emperor Tenmu[7] took the decision to build a shrine for Amaterasu that would be kept from one festival to the other. Among the impressive chain of ceremonies occasioned by the Jingu Shikinen Sengu festival that is taking place every 20 years, an essential moment was the consecration of the new sanctuary for Amaterasu and the beginning of the dismemberment of the 20 year old one (Fig. 3.3).

---

[5] Japanese deity.

[6] Kamis represent the force of nature and derive from it; therefore the connection with the human life and the influence and interaction are obvious.

[7] Late seventh century.

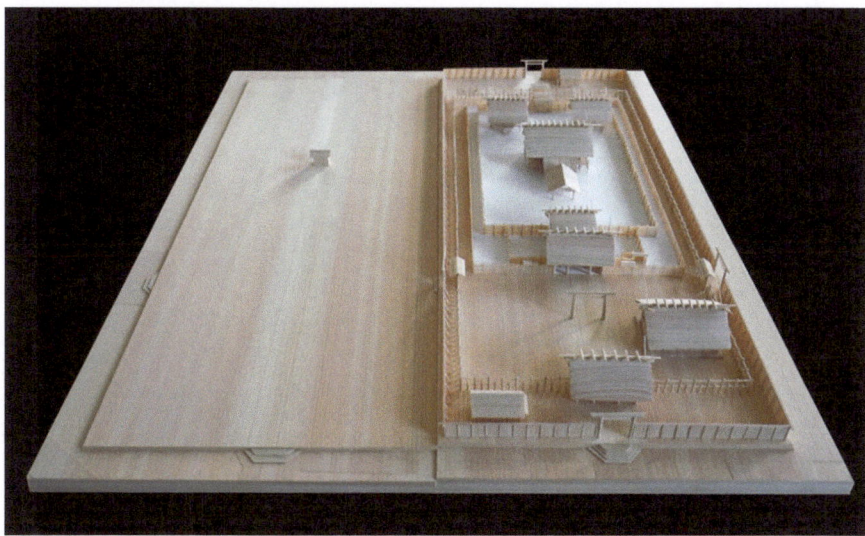

Fig. 3.3 Ise Jingu Shinto shrine (Fukami exhibition, Paris) 1:200 model of the Ise Shinto shrine: Ise Jingu. Model presented in the exhibition "Fukami. A dive into Japanese aesthetics". concept: Masahiko Tsugawa curator: Yuko Hasegawa. (Photo Jean-Pierre Dalbéra, CC BY 2.0 DEED. Source: https://www.flickr.com/photos/dalbera/42603600380)

In time, the outer shrine (home of Toyouke, the kami of agriculture and industries and Amaterasu's companion and food provider) as well as the Tori Gate and the Uji Bridge over the Isuzu river were also included along with the most revered place, Amaterasu's shrine, in the rituals of the periodical re-erection of the constructions.

2013 was the year of the 62-nd reiteration of the Shikinen Sengu festival.

The Shinto philosophy behind the periodical rebuilding of the edifices deals with the deep respect and understanding of the nature: renewal follows death; day follows night; spring follows winter; the circle closes and a new course begins. Connecting to the natural cycle boosts the energy in the sacred places. It is the awareness of impermanence that repeats endlessly the cycle of life and death. The high respect of the Japanese culture (and religion) for nature is the way the buildings are integrated in the landscape, through materials and scale, as a part of nature, in a substantiated dialogue. No dissonant shapes or materials that alter the spirit of the natural space, no disrespectful attitude of the built environment over the natural one, the two being parts of the same whole, which in fact **is** the environment (Fig. 3.4).

Fig. 3.4 Ise Jingu Shrine: walkway that descends from the Amaterasu shrine(s)—left and walkway towards the Uji Bridge—right. (Photo: Ana-Maria Dabija)

To make sure that the building technologies were preserved and passed without changes from one generation to another, almost as a part of the whole ritual, the 20 years interval gives the possibility for learning, exercising, and teaching the specific techniques. This architectural style[8] may not be used elsewhere [4]. The "twin sites" were decided more than a millennia ago and, with very few delays, the building, dismantling, and rebuilding took place despite wars and troubled periods (Fig. 3.3). The building(s) take(s) 8 years to complete (4 years only to prepare the wood), according to precise timing and ceremonies and for a short while, the two identical shrines—the old and the new—stand side by side. The most sacred objects of worship are transferred to the new shrine and in the old one the dismantling process begins. Forty-three of the other structures are also rebuilt every 40 years. The buildings are made of 12,000 logs that originate from 200-year-old cypress trees, using traditional tools, wooden dowels and interlocking joints (as do most old wooden houses throughout the world).

---

[8] Yuitsu-shinmei-zukuri.

The wood originating from the dismantled shrine is used to build a new Torii (entrance) gate or is incorporated in other shrines, to re-energize them. Sacred objects are also made and distributed to shrines across Japan.

On the other hand, there is a very down-to-earth spirit in this periodical building restatement: Japan is a country with high seismic hazard; a major earthquake is bound to happen in the interval of 20 years. Therefore it is maybe more efficient to operate with ever-new buildings instead of repairing them, especially when the building is constructed with components that can be dismantled and reassembled.

Although the idea was to preserve the ancient techniques due to religious reasons, the fact that they were transmitted from one generation to the next is an extraordinary scientific and cultural gain: the preservation of the techniques, from the tools to the smallest detail, for almost two millennia is like living archeology. Even if the architectural style of the shrine is unique as mentioned before, some things are redundant: the climate, the building materials that, together, lead to the same technological details. The pedagogical method is also worth mentioning: an apprentice would assist the masters in the first iteration when he observes and learns the skill. After evolving professionally during the next 20 years, he has the opportunity to maturely understand and apply the specific technology (under the supervision of a more experienced artisan and while a next generation of apprentices watches and helps) and in the final "loop" (or sequence) he can pass the know-how over, as a master this time, to another generation of apprentices. In other words, in a period of 40 years a trained worker could perform a truly mature knowledge transfer.

The millenary example of the Shrine in Ise as well as the timeless vernacular architecture[9] prove one thing: once the system is established, there are no fundamental changes to it. Nobody questions the principles; the work is done correctly and the theory has no involvement in the work being done.

And what is the relevance of the story of the Ise Shrine for the subject of energy efficiency in buildings? Actually, buildings are relevant for nearly any subject, as most of our activities are carried out in buildings. And in this specific case, the fact that the building components are integrated in other buildings or are used for entirely new constructions is an approach that shows respect for the environment and reduces the process of depleting the natural resources (or, according to other value scales, diminishes the carbon footprint). The idea of distributing building components and objects as artifacts to other religious edifices aims to preserve and expand the energy of the holy objects; the same approach has a more practical dimension in the traditional architecture where the re-use of the building components is also a priority: why waste a good item that is less expensive than a new one and provides the same result? The goal was/is not to save the planet but to save the budget. And it worked for centuries.

Closing the loop, it works today as well: energy is saved as no embodied energy adds to the building materials and components.

---

[9] More on this topic in Chap. 4.

Recuperate-Repair-Reuse: From 3 to 7 "R"s

This approach was referred to, in the past two decades, as "circular economy."[10] It is based on three (common sense) principles [5]: eliminate waste and pollution, circulate products and materials, and regenerate nature. The principles of circular economy oppose to the linear development of "take-make-use-dispose" [6] and are included in strategies that aim to reuse, repair, and high-quality recycle the products, in order to decrease the use of raw materials on the one hand and to decrease the amount of waste on the other hand (Fig. 3.5).

Fig. 3.5 Plastic waste. (Photo: Ivan Radic CC BY 2.0 DEED. Source: https://www.flickr.com/photos/26344495@N05/51212949244)

Probably the principles of repair-reuse-recycle can be applied in all the fields of human activity but our focus are the building-related products.

The abovementioned documents [6, page 11] state, in what constructions are concerned that "The built environment has a significant impact on many sectors of the economy, on local jobs and quality of life. It requires vast amounts of resources and accounts for about 50% of all extracted material. The construction sector is

---

[10] More on this topic in Chap. 2.

responsible for over 35% of the EU's total waste generation. Greenhouse gas (GHG) emissions from material extraction, manufacturing of construction products, construction and renovation of buildings are estimated at 5–12% of total national GHG emissions." The European Commission undertook in 2020 *The Circular Economy Principles for building design,* [7] a document[11] that aims to provide "principles for circular design of buildings" and to address all actors "involved in the construction industry, including economic operators in the value chain, policy-makers, legal and technical." The aim of the document is to contribute to the improvement of sustainability in buildings, emphasizing "the necessity of the reduction of waste, the optimisation of material use, and the reduction of environmental impacts of designs and material choices throughout the life cycle" through the following objectives:

1. *"Durability: building and elemental service life planning, encouraging a medium to long-term focus on the design life of major building elements, as well as their associated maintenance and replacement cycles;*
2. *Adaptability: to extend the service life of the building as a whole, either by facilitating the continuation of the intended use or through possible future changes in use—with a focus on replacement and refurbishment;*
3. *Reduce waste and facilitate high-quality waste management: facilitate the future circular use of building elements, components and parts, with a focus on producing less waste and on the potential for the reuse, or high-quality recycling of major building elements following deconstruction. This includes efforts along the value chain to promote: (1) the reuse or recycling of resources (i.e. materials) in a way that most of the material's value is retained and recovered at the end of a building's life span; (2) the component design and the use of different construction methods to influence the recovery for reuse or recycling to avoid down-cycling".*

As in many other cases, one can observe that the documents and regulations only confirm (in many words and many pages) what the living beings have been doing throughout history: they have built efficiently, according to the available materials and technologies of their time and shaped the space according to the structural, functional, and aesthetic requirements that existed.

Today, for different reasons than a century or a millennia ago, the idea of recuperating, repairing, and recycling is, at least in theory, present in most aspects of our lives: home appliances are rarely repaired by their owners and the "buyback" repurchasing system can be regarded as a type of circular economy.

The most common approaches to efficiently use existing materials and components is to recycle, repair, and reuse: an attitude that has a major impact on the energy saving process.

---

[11] Elaborated by a large group of stakeholders that include representatives of the whole professional field: standardization, building materials manufacturers, confederations and associations of builders, architects, and surveyors

Primum Non Nocere (First Do No Harm)

Prior to establishing if—and how—specific materials should be recycled, there are two preliminary steps that must be undertaken:

- An energy-use analysis, to assess whether the effort of recycling does not use more energy than the one of producing goods with virgin raw material
- A health hazard analysis, to assess whether the recycled product is not toxic for the living world and, hence, for the planet

Maybe the highest percentage of waste in today's world consists of plastic. In the context of the pressure put during the pandemic of 2020–2023 by the need of increased hygiene, the industry of single-use plastic products skyrocketed.

According to Shan et al. [8], "a 2021 exploratory study observing 202 households from 41 countries found a 50% increase in food packaging and a 35% increase in single-use bags in households' plastic waste."

Plastic can be recycled mechanically, chemically, and thermally [9]. Of the three methods, the most commonly used is the mechanical method (material recycling) but it has the disadvantage of a low-quality final product, obtained with higher costs (compared with the new/virgin plastic) and with high NOx emissions. The chemical recycling has a higher greenhouse gas footprint than the mechanical one. The thermal recycling seems to have several disadvantages (high greenhouse gas footprint, toxicity, given by the emission of dioxine) and is banned in some countries (like Japan, as we referred to the [8] research).

Overemphasizing the possibilities of reintegrating the original materials in new products is advantageous for the image on the company or for the generic product but not always for the health of the living beings so the idea of "saving the planet" through recycling is somehow exaggerated as the recycled product may be toxic to animals—humans included.[12]

It is the case of plastics, recent research [10] point out that "When scientists examined pellets from recycled plastic collected in 13 countries they found hundreds of toxic chemicals, including pesticides and pharmaceuticals. Because of this, the scientists judge recycled plastics unfit for most purposes and a hinder in the attempts to create a circular economy."

Scientists from the University of Gothenburg, IPEN, Aarhus University, and the University of Exeter concluded that: "The hazardous chemicals present risks to recycling workers and consumers, as well as to the wider society and environment. Before recycling can contribute to tackling the plastics pollution crisis, the plastics industry must limit hazardous chemicals." "Thermal recycling should be used sparingly, as burning plastic results in harmful pollutants such as dioxin" [8].

---

[12] In the late 1980s and early 1990s, when asbestos-cement was discovered to be a hazardous material, some products mentioned on their labels that they "do not contain asbestos," without indicating what they did contain; some researches proved that the new content was also hazardous for human health.

In other words, the aim should not be to recycle at any costs but to find the appropriate field of application for the disposed products that would diminish the heaps of waste, would decrease the costs of the building assemblies, and would be safe for the whole natural environment as well, with all its living beings.

In this respect, the famous *First do no harm* attributed to Hippocrates applies in the building industry as well (and probably in most fields of activity); the waste exists, it is produced by humans and while some of it is safe—like wood, clay, glass, or metal—and can be recycled without threatening human health, the situation is different where plastic is concerned.

Maybe finding possibilities for integrating the plastic bottles into building components could be a better idea, not for reducing the carbon footprint (which they, in fact, do) but for reducing the risks of ingestion of toxins and for diminishing overall waste generated by plastic. Obviously, such an approach should be preceded by scientific research, linked to the geo-climatic conditions of the site. Hence, methods, calculations, simulations, tests, prerequisites, special provisions, and an entire philosophy that frames the building component(s) in classes and provides theoretical and practical information regarding the field of application are gathered in regulations that back up an innovative structure. As seen in Fig. 3.6 building walls and fences can be made with plastic vessels (mainly as demonstrations).

Fig. 3.6 La casa ecológica de botellas. Misiones Argentina. (Photo: Rodrigo Soldon CC BY-ND 2.0 in Source: https://www.flickr.com/photos/soldon/5684857431)

A more feasible approach is to transform waste materials into sculptural elements (Fig. 3.7), enriching public spaces in the urban area. It is also true that transforming global plastic waste into urban sculptures is not the point we want to make (and these sculptures do have an end of life themselves) but still, here and there we consider it is worth the effort.

Fig. 3.7 Chameleon sculpture, made of waste bottle caps. (Photo: Becks, CC BY 2.0, in Source: https://www.flickr.com/photos/littlemisspurps/9117175401)

Closing the Loop

In very simple words, "circular economy" can be defined as the art of saving: nothing is lost, everything is transformed and therefore costs are minimized to a maximum; once a product ceases to be used—for one reason or another—a new field of application is found for it. And the loop closes and the journey starts again.

This approach can be illustrated by the story of straw constructions (not a new idea either): straws are by-products of cereal plants. Their applicability in the building sector has been speculated throughout the whole history of buildings, as roofing material—thatch—or as wall building component, bale. In the mid-nineteenth century, with the invention of the mechanical hay baler, the hay packages began to be used in the construction of the walls of the buildings in America. As many other inventions, it eventually was abandoned but entered in the cone of interest of the specialists in the late twentieth—early twenty-first century, as an interesting and ecological option for thermal insulated roofs and walls, while reducing the amount of wasted hay that was exceeding the capacity of the landfill space (even though entirely biodegradable). Obviously, it is also saving energy, as the thick by-product is an efficient thermal insulation. In order to use it safely (the material is not fire resistant and it is sensitive to insect contamination and attack), some countries[13] carried out regulations that provided the conditions of use and the limitation of these building products. Today these building systems are not used only as thermal-efficient systems but they often emphasize unique architectural characteristics of the building(s), of the area(s) or of historic and cultural periods.

A good example is the case of The Shakespeare Globe Theatre, in London (Fig. 3.8). The idea was tackled for more than two decades and the theater finally opened its gates in 1997. It was designed as a replica (which it is not and cannot be, as nobody knows precisely how the original Globe Theatre was) that used the sixteenth-century building know-how and displayed the first—and only—thatched roof in London since the Great Fire of 1666.

Fig. 3.8 Shakespeare's Globe, London. (Photo: City.and. Color, CC BY 2.0. Source: https://www.flickr.com/photos/technicolourcity/13976124258)

The original Globe was built in 1599 with a timber structure entirely recuperated from a previous theater construction: The Blackfriars Theatre. This sounds familiar: recuperating-refurbishing-reusing, but unlike the case of the Great Shrine, moving the structure from one location to another did not hold any spiritual or emotional significance although there was a functional continuity of the two buildings: both The Globe and The Blackfriars were playhouses. During its 45 years of functioning

---

[13] USA, some European countries, Australia, and maybe others as well.

(it was demolished in 1644 [11]), it caught fire, was reconstructed, was closed, and eventually was pulled down. According to Nagler [12], on June 29, 1613 the theater burned down in 2 h, during a representation of Henry VIII due to a cannon salute fire "as the King entered the Cardinal's house and a burning wad of paper stuck in the thatched roof. There seems to have been no panic or loss of life. One man's breeches took fire but this minor blaze was extinguished with a bottle of ale. The event was celebrated in a number of popular ballads, one of which advised actors to whore less and spend the money thus saved to construct a tile roof."

As Shakespeare's Globe Theatre had to respect the contemporary building codes and regulations, adjustments had to be made in order to fulfill the necessary requirements mainly on fire safety. Hence, the thatched roof withstood a chemical treatment, consisting of a mixture of borax and boracic acid sprayed on and adhered with a PVC adhesive. Active measures for fire safety were adopted by providing specialist sprinkler system and 600 liters of fire retardant fluid [13]. Obviously these provisions raised the cost of the building but, as previously mentioned, the aim was not to make a demonstration of implementation the principles of circular economy or energy efficiency but to create a space with the atmosphere as close to the Shakespearean era as possible.

Wood

Although a material that can easily be processed, wood does not have a tradition in contemporary recycling—at least according to the National Waste Associates [14]. According to Carmine Esposito "After concrete, wood is the second greatest component of construction and demolition (C&D) waste. Nationwide, wood contributes between 20–30% of all C&D related debris and it accounts for almost 10% of all material sent to landfills each year."

The numbers are daunting, as wood is relatively easy to handle: it can (almost) always be scrapped as it is (almost always) installed by mechanical means, which allows un-screwing the parts and bringing them down with a minimum of effort or loss. In a similar way, the parts can be transported and re-installed identical or not on a different site. Wood can be cut, shaped, grounded, transformed, and integrated in different products; the wooden sub-assemblies can originate from windows, doors, shutters, frames, fences, and floors, from products that used waste wood already (pallets, chipboards, and fiberboards) but "unlike concrete and steel, which have relatively high recycling rates (98% and 82% respectively), lots of C&D derived wood waste goes straight into the dumpster and then to landfill, costing construction companies heavily. [...] And worse still, according to a U.K. study, as much as 10–15% of wood procured for new construction projects ends up in recycling or waste streams without being used at all." [14].

Approaches like the one in 1599 of dismantling and reassembling a wooden structure into a new building or like the one that ritually takes place every 20 years, for the past fourteen centuries, in Japan did not become a rule, although there are some beautiful examples of integrating existing wooden components in new buildings.

The Kamikatz Public House (also known as The Kamikatsu Zero-waste Center or the "WHY"), designed by Hiroshi Nakamura and NAP, opened in 2020. This statement-building is located in Kamikatsu, a "zero waste" town in Tokushima prefecture where 80% of its waste is sorted into 45 categories and recycled. The facades display doors, windows, and shutters recuperated from abandoned houses (Fig. 3.9). Used items are exposed at the recycle center.

Fig. 3.9 Kamikatsu Zero Waste Center. (Photo author: unknown. Public domain—CC0. Source: https://en.wikipedia.org/wiki/Kamikatsu_Zero-waste_Center#/media/File:Kamikatsucho-zero-waste-center.jpeg)

Another example of a statement-building is the Collage House, in India designed by S+PS Architects (Fig. 3.10). The name of the building is emblematic: its main facade is, indeed, a collage of recuperated shutters, windows, and doors of different shapes and sizes, fit together into an assembly that achieves a balanced contrast between the old and the contemporary building materials—especially apparent concrete, glass, and polycarbonate.

Fig. 3.10 Collage House, in Mumbai, India, designed by S + PS Architects. (Photo: Photographix India. Courtesy S+PS Architects)

Furthermore, all the reclaimed shutters preserved the hardware and are fully functioning, allowing for the natural ventilation of the interior spaces, hence providing continuity not only with the old building elements (windows, shutters, and doors) but with the Indian traditional building systems where the interior air quality and comfort were accomplished by the deep knowledge of the rules of nature.[14] And of course, saving at least the embodied energy of the reclaimed building components.

Metal

Metal can be recycled almost 100% over and over again, once it is separated in individual components of the same type of material. However, this is easier said than done, due to several reasons and circumstances: the process requires specialized management and equipment for the completion of defined stages—separating ferrous metals (iron, steel) from non-ferrous metals (aluminum, brass, copper, zinc, etc.), crushing, compacting, smelting (which in turn is dependent of the melting point of each metal or alloy that has to be processed as well as on the size of the furnace), and removing the contaminants. The number of metals in our everyday basic products exceeds 45.

More recent policies in the field recommend to optimize the recycling of entire product that reaches its end of life instead of focusing on extracting the components

---

[14] More on this topic in Chap. 4.

and recycle them individually. Table 3.1 presents a synthesis of the facts declared by EuRIC [15] for the most used metals: steel, aluminum, and copper.

Table 3.1 Synthesis of the advantages of recycling metals, from the environmental perspective, according to [15] The data refers to 2017–2018

	Saves energy for primary production	1 ton of recycled material saves virgin material	Air pollution	Water pollution	Water use
Steel	72%	1.4 tons of iron ore 0.8 tons of coal 0.3 tons of limestone	Reduced by 86%	Reduced by 76%	Reduced by 40%
Aluminum	95%	8 tons of bauxite 7.6 cubic meters of landfill			
Copper	85%	NA			

According to IRP-UNEP[15] Report of 2013 [16]

"While common commodity metals like steel, magnesium and copper can be recovered relatively easily, as these are often used in relatively simple applications, the small amounts of metals in, for example, electrical and electronic waste can be harder to recover because they are often just one among up to 50 elements. As an example, a mobile phone can contain more than 40 elements including base metals such as copper and tin, special metals such as cobalt, indium and antimony, and precious and platinum-group metals including silver, gold, palladium, tungsten and yttrium. Fluorescent lamps contain various materials and elements which include a range of Rare Earth elements, and other critical metal resources. And a modern car contains nearly all metals available, as it is a product that integrates a broad range of other metal-containing products."

The process of smelting the individual scraps and alloys is—obviously—energy consuming (as are all the metallurgical processes) but the gain is reducing the waste and the depleting of virgin resources, as well as saving energy from the mining activity. Therefore, a "linear, one-dimensional, approach cannot deal with the complexity of interactions between metals, mixing of waste streams, and the economics behind processes. [...] Where the economic incentives for collection of waste by private or public operators are not aligned with policy goals, significant resource volumes can be lost to illegal or informal recycling, or are simply unaccounted for, sometimes through 'cherry picking'. This often leads to environmental problems, damaging health impacts and impacts on water or climate, as regulatory standards are ignored" [16 page 34]. In other words, recycling metal is almost 100% efficient provided that it is doubled by corresponding policies and economic approaches, which, in fact, means that the bottom line is that the action is expensive, energy consuming, and sometimes hazardous for the humans.

Using components and pre-existing assemblies whenever possible is a good idea and a respectful attitude toward the environment, in the urban landscape (as seen in Fig. 3.11), or in the building design (see Fig. 3.12) but it does not fit in the specific location.

---

[15] International Resource Panel, United Nations Environment Programme.

Fig. 3.11 MetalScapes by S.W. Huffman from Ottumwa, Iowa. (Photo Carol VanHook, CC BY-SA 2.0. Source: https://www.flickr.com/photos/librariesrock/28288815303)

Fig. 3.12 World Cup 974 Stadium in Doha. (Photo: Michael Coghlan, CC BY-SA 2.0. Source: https://www.flickr.com/photos/mikecogh/52947158581)

Easier said than done, at least where a building design is at stake but, as the example of the Doha World Cup 974 Stadium shows, "when there is a will, there is a way."

Located on the waterfront and displaying the spectacular view of the Doha skyline, the 974 Stadium aimed to replicate/echo /mirror the industrial area of the harbor.

Besides, the concept took into consideration that several huge buildings, designed and constructed for targeted events, lost their utility once the event was over.[16] Therefore, Fenwick Iribarren Architects conceived an arena that could be demountable, transportable, and reusable elsewhere [17, 18].

The structure consists of a steel frame that accommodates 974 different and colorful certified shipping containers, hence transportation by sea from one location to another can be safely managed. All the facilities (sanitation included) are integrated in the blocks that assemble/disassemble like logo pieces. The capacity of the stadium is 44,089 spectators.

The stadium functioned for the 2022 World Cup, hosting seven matches, from 30 November 2021 to 5 December 2022 and the process of dismemberment was due to begin after the FIFA competition ended. However, so far no decision regarding the new location was taken and in March 2023 the stadium was still standing [19].

According to Associated Press [20]

> "Carbon Market Watch, an environmental watchdog group that investigated Qatar's World Cup sustainability plans, said whether Stadium 974 has a lower carbon footprint than a permanent one comes down to "how many times, and how far, the stadium is transported and reassembled." FIFA and Qatar acknowledge that in a report estimating the stadium's emissions. If the stadium is reused only once, they estimate its emissions would be lower than a permanent one as long as it is shipped fewer than 7000 kilometers (about 4350 miles) away. If it's repurposed more than once, it could be shipped farther and still be less polluting than a permanent venue, they said, because of how energy-intensive building multiple new stadiums is."

Ceramics

As surprising as it may seem, the recycling of ceramics is extremely nuanced and complicated: not all ceramic waste can be recuperated by crushing and remelting—as do metal or glass waste—due to recycling issues/hazards, energy consumption/ emissions, toxic chemicals, and lead exposure [21]. Ceramic production is energy intensive, involving very high temperatures for the baking and firing processes. Besides, during the process it produces significant toxic emissions like particulate matter, nitrogen oxides, carbon monoxide, carbon dioxide, sulfur dioxide, hydrogen fluoride, VOCs, and heavy metals, posing a health hazard on the workers as well. Accidentally mixing broken ceramics with glass cullets and smelting them in glass

---

[16] The Millennium Dome in London, for instance, housed an unsuccessfull exhibition throughout 2000 and, once the exhibition was over, found with difficulty an appropriate architectural function. It closed down in 2001 and was subject to renovation that transformed it 7 years later from an exhibition to a 20,000 seats auditorium [22].

furnaces may destroy the glass recycling equipment, as the melting temperature of ceramics is significantly higher than the melting temperature of glass, respectively, around 2000 °C.

Crushing waste tiles or bricks and integrating them, as fine clay or rubble in new mixtures and structures, diverts the waste from landfill (with all the known advantages, from reducing the demand of the specific natural resource to a better management of the landfill). However, these technologies seem to be still in experimental stage.

Recycling ceramics is more an approach of finding new fields of applications for the existing products: more than a century ago, Gaudi[17] used a special technique of tiles and mosaics made of colored ceramic crocks (trencadís) for finishing surfaces of buildings and exterior furniture. The exterior benches (Fig. 3.13) and sculptures as well as the walls and ceilings of the buildings within the Park Güell in Barcelona are covered with *trencadis* not for reducing the ceramic waste or diminishing the carbon footprint but because the colored ceramic was too beautiful to be wasted. As previously stated, it is up to the architect's imagination to use existing products and integrate them in the building, when appropriate. The bonus is that the overall energy required in the endeavor of finalizing a building, from the drawing board to the actual construction decreases with the embodied energy of the components (and with the shaping of the space that takes nature into account[18]).

Fig. 3.13 Bench in the Park Güell, Barcelona, Spain. Architect Antoni Gaudi (1852–1926). (Photo: Ana-Maria Dabija)

---

[17] Architect Antoni Placid Gaudí i Cornet (1852–1926), prominent figure of the Spanish Art Nouveau style.

[18] More on this topic in Chap. 4.

The bricks or the roof tiles of the demolished buildings can be evaluated and integrated in other buildings of the same epoch, accomplishing a unitary intervention (rather than using new and antiquated building materials with a different mechanical strength and hygro-thermal properties but with higher embodied energy). The process of cleaning and verifying the bricks and tiles of old, degraded buildings may be time consuming but it is worth the effort, as often the mechanical resistance of the recuperated building elements is higher than the one of the contemporary products, not to mention that, as the brick or tile dimension and color varied in time, interventions over existing buildings can easier be made with materials of the same period.

Special projects can be carried out, to preserve, present, and emphasize traditional characteristics of a specific culture. It is the case of The Porcelain House in Tianjin, China (Fig. 3.14) which exposes "over 400 million pieces of ancient Chinese ceramic chips, more than 5000 ancient porcelain vases, 4000 porcelain plates and bowls, as well as over 400 white marble carvings, 20 tons of natural crystals and agates, and over 300 stone lions in diverse sizes of different times" [23], bringing to light porcelain from Song dynasty,[19] Yuan dynasty,[20] Ming dynasty,[21] and Quing dynasty.[22] It took artisan antiquarian Zhang Lianzhi over 20 years to collect materials and 5 years for the decoration.

---

[19] From 960 AD to 1279 AD.
[20] From 1271 AD to 1368 AD.
[21] From 1368 AD to 1644 AD.
[22] From 1644 AD to 1912 AD.

Fig. 3.14 The China House in Tianjin, China. (Photo: Steve Knight, CC BY 2.0. Source https://www.flickr.com/photos/kitmasterbloke/17890812988)

Glass

The story of man-produced glass goes back in history about at least five thousand years. A romantic legend about the accidental discovery of glass recounts that a Phoenician commercial ship landed somewhere on the beaches of Syria. As it was evening, the crew went on the shore to cook their meal and, as they did not put out the fire under the pots, by morning the sand melted and turned into a vitreous mass. Specialists say that this is only a legend, as the temperature needed for the silica—the sand—to melt is by far higher than the one reached by the fire under the pots [24]. The reality is that, discovered by accident or not, glass is one of the most precious materials of the Antiquity [25]. Glass consists of silica (sand), lime, and soda ash, provided in specific proportions. The compound melts at 1300 °C and turns into an amorphous mass. In its journey of more than five millennia of use, it became, from a luxury item/accessory, one of the most common and domestic objects surrounding us (although it should be mentioned that glass is still one of the most versatile materials with a dynamic evolution and a yet not fully discovered potential).

As it consists of natural ingredients, it can be recycled, by smelting in furnaces at high temperatures. According to [26], "in 2018, 31.3% of glass food and beverage packaging containers were recycled. In some states, like California, glass bottle recycling reaches over 80%." The same source provides the following facts on saving the natural, raw resources: "over a ton of natural resources are conserved for every ton of glass recycled, including 1300 pounds of sand, 410 pounds of soda ash, 380 pounds of limestone, and 160 pounds of feldspar." By using glass cullets in the manufacturing process of glass, a decrease in energy use is registered (2–3% drop with every 10% cullets, according to the same source).

Melting and recirculating glass—using higher or lower percentages of reclaimed material in the virgin mixture—reduces the use of energy and these savings can be emphasized in the (more or less) embodied energy of the final product. Still, this is not the aim of our endeavor as much as the idea that finite products can be also employed as building components, hence cancelling the costs (physical as well as in terms of the use of energy) of the processing of the existing items within the new products.

Examples of bottle houses where a minimal intervention on the waste is applied are referred to mostly as curiosities or excentricities than as substantiated approaches (Fig. 3.15).

Fig. 3.15 Waste bottle walls: glass bottles chapel wall in Canada. (Photo: Cara Fealy Choate, CC BY 2.0, in Source https://www.flickr.com/photos/carabou/9257040946)

Obviously it is not always possible and we don't intend to make generalizations but while most glass is recycled by the glass producers, containers may sometime be expressive materials in the hands of the designer to contribute to the decreasing of waste and to minimize the embodied energy in some (few) buildings.

Circular Economy Applied: Building Conversions

A functioning building that has come a long way of hundreds of years, having the original function or a different one, is the perfect expression of resilience: it shows adaptation in time to the impact of all categories of agents.

Older buildings typically use more energy than new ones [27], just as old equipment uses more energy than the contemporary "A+++" household devices. Therefore, while the shell often remains, the interior space needs to be adapted to the periods' requirements. The conclusions may be that our predecessors built durable constructions indeed: we can still see them, admire them, learn from them, and use them. Throughout time, the buildings withstood architectural conversions (changes of functionality) in order to adapt them to the epochs' demands.

As the society changes, the cultural, economical, or technical perspective changes as well while the shell often remains as a memento of a past era. The buildings become outdated not due to stone, bricks, and mortar but to the evolution of the society; in other words when, for some reason, the architectural program changed or disappeared and the buildings lose the scope for which they were erected in the first place. It is often the case of industrial buildings or industrial sites. When this situation occurs, one of two positions need to be adopted:

- To demolish the building (with all the costs and consequences that derive from this action)
- To study, adapt and use the building, giving it a new life by accommodating a "contemporary" architectural program within the existing shell. The term should be considered in a temporal dimension: many museums, for instance, were organized in former palace buildings. The best known example is the Louvre, residence for French kings in the sixteenth century, transformed into the building that hosted the private royal collections in the seventeenth century, transformed, and used as a Royal Academy for a century and finally turned into a public museum by the French Revolution.

The idea of conversion was not—and should not be—related to the reduction of the carbon footprint of the buildings or to the circular economy principles; it is a *common sense* attitude, of saving time and money, in order to accomplish something—in the given case, to accommodate a set of architectural functions in a given space, with the minimum of disturbance.

Besides, occupying abandoned built spaces is a process adopted by most of the living creatures: as mentioned before, it is the logical and easier thing to do.

Without pretending that these interventions have saved the planet, we must nevertheless note that the concern to bring a new life into existing buildings dates back longer than initiatives on the sustainability and resilience of construction. Moreover, a building that crosses hundreds of years and is in operation—be it the original one or another—expresses the idea of resilience: adaptation to the impact of agents (of all categories), over time. And in the case of converting existing buildings from one function to another, the embodied energy[23] will only refer to the new interventions—in other words, it is significantly reduced. Undoubtedly, the shell might be preserved (partly or totally) but the structural system would need to be rethought according to the new spatial demands, as would parts of the horizontal (floors) and vertical (walls) partitions. New facilities and equipment will need to be accommodated, in accordance with the new configuration and architectural function.

During the twentieth century, building conversions have been carried out more frequently as the building equipment and installations evolved more rapidly.

Perhaps the most plastic examples from this perspective are the abandoned Gothic churches, in Western Europe, which have been adapted for use in recent years, shifting from the program of religious buildings to a large range of programs of civil architecture: from housing to clubs, from agro-food markets or public food spaces (Fig. 3.16) to libraries, from museums to hotels. One example is a food market in London, located in a nineteenth-century church that was transformed into antiques market, after being deconsecrated in 1974. Recently it was subject of a reconversion, from commercial to alimentation function.

---

[23] The embodied energy is associated with the building process; it is the energy that is required to construct something, from the stage of obtaining the raw material to the final phase, when the product is finished and ready to use. The term has been further nuanced [28] and it was given an economical connotation, in approaching rehabilitation or/and building materials originating from demolition/dismantling by association with the sunk costs: it cannot be recuperated as it was already "counted" when the building was first constructed.

Fig. 3.16 Nave of Mercato Metropolitano, Mayfair. (Photo: Matt Brown, CC BY 2.0 DEED in Source: https://www.flickr.com/photos/londonmatt/49131693741)

In other cases the changes of the technology made some buildings obsolete. Two famous examples are the Vienna Gasometers and the Musee d'Orsay in Paris.

The four gasometers, located in the outskirts of Vienna, were built at the end of the nineteenth century with the purpose of storing the gas reserves, necessary for the functioning of the city (Fig. 3.17). In 1984, by switching to the use of natural gas, the storage method also changed and, as a result, the huge masonry structures—each gasometer having a height of 75 m and a diameter of 60 m, ensuring a storage capacity of 90,000m^3—were abandoned. However, they were not demolished and were eventually included in the list of historical monuments.

Fig. 3.17 The four Gasometers. (Photo: Abductit CC BY 2.0 Source: https://www.flickr.com/photos/42231620@N07/4220513493)

In 1995 Vienna municipality called for ideas for integrating these monumental structures in the urban tissue, within a greater project of the revitalization and remodeling of the urban area of which they were a part.

Each of the four buildings was assigned, for the conversion, to an important name in contemporary architecture: Jean Nouvel, Coop Himmelblau, Manfred Wehdorn, and Wilhelm Holzbauer.

The common function for all the four conversions is the residential architectural program but within the different constructions it combines with office, entertainment spaces, or retail (located on the lower floors and connecting all the four structures).

Identical on the exterior, each gasometer emphasizes a different interior space, according to the architect's vision and to the detailed architectural theme. While the 14 levels of apartments of the Gasometer A, designed by Jean Nouvel (Fig. 3.18-left), are provided with large windows that catch the light that enters through the immense glass dome, the apartments of Gasometer D, designed by Wilhelm Holzbauer (Fig. 3.18-right), benefit of a beautiful inner garden.

Fig. 3.18 Interior space of Gasometer A, conversion by arch. Jean Nouvel (left). Gasometer D, conversion by Wilhelm Holzbauer (right). (Photo left: Abductit CC BY 2.0. Source: https://www.flickr.com/photos/42231620@N07/4220513387. Photo right: Andreas Pöschek, CC-BY-SA-2.0-at Source: https://it.wikipedia.org/wiki/File:Gasometer-d-by_viennaphoto_at.jpg)

Another famous example of conversion was the Orsay train station in Paris transformed (eventually) into a museum. Designed by Victor Laloux, Lucien Magne, and Émile Bénard and built in 1900 on the ruins of the Orsay palace [29], the proximity to the Louvre and the Legion of Honor represented an urban architectural challenge at the end of the century.

The Paris station (Fig. 3.19) operated, alongside the hotel of the same name, until 1939 when it became a hub serving the suburbs as a result of the electrification of the railways. The modern, much longer trains could not be served by the short platforms of the existing station that could not be extended. During the Second World War the station was used as a postal center, from where packages were sent to prisoners; it was turned into a film set; then it represented a temporary space for the Renaud-Barrault theater company, after which it hosted the Drouot auction house during the reconstruction of its own headquarters. In 1970 a permit for demolition was granted, to make room for the construction of a convention center and hotel [30]. The building was saved by one of the French ministers, who reversed the permit and managed to include the Gare on the list of historic monuments, in 1973. By 1975 (and with a final approval in 1977), as the interest in the art and architecture of the nineteenth century increased, the decision was taken to adapt the space of the railway station for the function of a museum where the

French art from the period between the classicism of the Louvre and the modern art of Centre Pompidou would be permanently exhibited. In other words, from the period between 1848 and 1914. Eventually, it turned out to be the most important museum on Impressionism.

Fig. 3.19 Orsay Station, Paris, Paris-Orléans railway. (Photo: public domain (postcard) CC BY-SA 3.0. Source: https://commons.wikimedia.org/wiki/File:Gare-d%27Orsay-BaS.jpg)

Subject of an architectural competition, the Musée d'Orsay (Fig. 3.20) was remodeled by A.C.T. Architecture, respectively by architects Renaud Bardon, Pierre Colboc, Jean-Paul Philippon. It opened its doors in 1986.

Fig. 3.20 Musee D'Orsay, Paris. (Photo: Dr. Bob Hall CC BY-SA 2.0 DEED. Source: https://www.flickr.com/photos/houseofhall/8414359797)

The museum was designed on three levels: on the ground floor, galleries are distributed on both sides of the central nave, the terraces of the median level overlook the nave and the top floor is located above the lobby. Appropriate services are spread throughout the three levels: the museum restaurant, the Café des Hauteurs, the bookshop, and the auditorium [31].

A Possible Conclusion

A possible conclusion would be: why not adapt and use what is there, instead of wasting materials, time, and money (which all can be transformed in energy) while preserving the history and culture in the same time (in a casual manner of speaking, this would mean hitting two birds with one stone).

However it must be mentioned that the professional practice—in other words, reality—has many nuances when it comes to take the decision of preserving or converting existing buildings. Sometimes it is less expensive to demolish and rebuild than to consolidate and adapt for a new function. Several owners of residential buildings too often want to demolish what they have and build anew. Using at least some principles of the contemporary current of circular economy contributes in a paradoxical way to the preserving of the built patrimony.

Besides, in the case of the existing built patrimony, it is not only the structure that needs to be strengthened; the building equipment is so dramatically different than what it used to be only 3–4 decades ago that in order to function according to the contemporary requirements, the architect must choose the appropriate compromise. Simply put, sometimes the costs (from spatial to economic) would surpass the principle of refurbish and reuse.[24] Of course this is also debatable, depending on what architectural program would the building be considered for. What we need to emphasize is that each situation needs to be dealt with as each building is unique, even if it is the result of a standardized design: the site, the orientation, the geographical position, and the local conditions—anthropic and natural—lead to changes or adaptations of the project and, just like in the case of living beings, houses are individuals and need to be treated accordingly and analyzed as living organisms.

References

1. https://www.isejingu.or.jp/en/about/index.html
2. Soul of Japan, An Introduction to Shinto and Ise Jingu. (2013). www.isejingu.or.jp. Accessed May 2023.
3. https://www.youtube.com/watch?v=_ttXjy_IZ6w
4. https://www.jrpass.com/blog/ise-grand-shrine-everything-you-need-to-know-about-japans-most-sacret-shinto-shrine
5. https://www.ellenmacarthurfoundation.org/topics/circular-economy-introduction/overview
6. Communication From The Commission to The European Parliament, The Council, The European economic and social committee and the committee of the regions. A new circular economy action plan for a cleaner and more competitive Europe COM/2020/98 final. https://eur-lex.europa.eu/legal-content/EN/TXT/?uri=CELEX:52020DC0098
7. https://ec.europa.eu/docsroom/documents/39984
8. Shan, C., Pandyaswargo, A. H., & Onoda, H. (2023). Environmental impact of plastic recycling in terms of energy consumption: A comparison of Japan's mechanical and chemical recycling technologies. *Energies, 16*, 2199. Academic Editor: Adriano Sacco, in. https://doi.org/10.3390/en16052199
9. https://thegreatbubblebarrier.com/plastic-recycling/
10. University of Gothenburg. (2023, November 10). Scientists found hundreds of toxic chemicals in recycled plastics. *ScienceDaily*. www.sciencedaily.com/releases/2023/11/231110112511.htm
11. https://www.britannica.com/topic/Globe-Theatre/The-design-of-the-Globe
12. Nagler, A. M. (1958). *Shakespeare's stage* (p. 8). Yale University Press.
13. Li, N. (2020, May 5). *Shakespeare's globe theatre*, Tuesday 14 June 2011 — Updated. https://www.facilitatemagazine.com/features/feature-articles/2011/06/14/shakespeares-globe-theatre
14. Esposito, C. (2021). *Wood recycling for the construction industry (Don't let your money get hauled away!)*. https://www.nationalwaste.com/blog/wood-recycling-for-the-construction-industry-dont-let-your-money-get-hauled-away/

---

[24] This is not happening only in the building industry; as previously seen, researches wave the red flags on the recycling of plastic, that is more toxic than the non-recycled one [32].

15. https://circulareconomy.europa.eu/platform/sites/default/files/euric_metal_recycling_fact-sheet.pdf
16. Reuter, M., Oyj, O., Hudson, C. DIW, van Schaik, A., Heiskanen, K., Meskers, C., & Christian Hagelüken, C.. *Metal recycling—Opportunities, limits, infrastructure, editor international resource panel*, Working Group on the Global Metal Flows, pp. 2–3.
17. https://www.sbp.de/en/project/stadium-974/
18. https://www.fifa.com/en/articles/stadium-974
19. https://www.theguardian.com/football/2023/mar/29/broken-promises-the-future-of-qatar-world-cup-stadiums-still-up-in-air
20. Built to Disappear: World Cup Stadium 974, (2022, December 3). https://www.voanews.com/a/built-to-disappear-world-cup-stadium-974-/6861177.html
21. https://zerowastewashington.org/ceramics/
22. https://www.britannica.com/topic/Millennium-Dome
23. https://www.visitourchina.com/tianjin/attraction/porcelain-house.html
24. https://epiphanyglass.com/history-of-glass/
25. https://whatson.cmog.org/exhibitions-galleries/origins-glassmaking
26. https://www.gpi.org/glass-recycling-facts
27. https://www.europarl.europa.eu/RegData/etudes/STUD/2016/587326/IPOL_STU   (2016). 587326_EN.pdf, p. 11, Accessed Sept 2021.
28. Williamson, T., & Pullen, S. (2009). *Sunk embodied energy as a means of valuing the built environment*, 43rd Annual conference of the architectural science association, ANZAScA, University of Tasmania. https://anzasca.net/wp-content/uploads/2015/12/ANZAScA_2009_Williamson_Pullen.pdf accessed nov 2023
29. https://www.musee-orsay.fr/en/collections/history-of-the-museum/the-station.html?S=0, Accessed September 2020.
30. https://www.parisinsidersguide.com/history-of-musee-d-orsay.html
31. https://museedeorsay.weebly.com/from-station-to-museum.html
32. https://www.theguardian.com/environment/2023/may/24/recycled-plastic-more-toxic-no-fix-pollution-greenpeace-warns

Chapter 4
Architectural Means and Tools for Providing Indoor Air Comfort: A Historic Perspective

Introduction

The passive design methods are based on the deep understanding of the rules of nature: passive gain and passive cooling are, according to the geographic area, specific means of providing indoor comfort conditions with the least energy consumption.

Buildings were kept warm or ventilated throughout the whole history of mankind; it is not today that we "invented the wheel." Today the goal is to provide better living conditions, with more performant systems. However, in order to use less resources and less fossil fuel, maybe looking back to what mankind did, in terms of providing indoor climate control, might be a good idea.

Good-quality air means controlling temperature, humidity, and air movement within the building. It is worth mentioning that passive means of design apply permanently, while the active means are triggered by pressing a button that puts into operation an equipment (that uses more or less conventional fuel). Passive design methods are based on the profound understanding of how nature works; active methods are based on the technological means and equipment and can be switched on and off by the user at any time.

It is also worth emphasizing that usually, in order to meet the contemporary demands in respect with the indoor air characteristics, both passive and active measures are adopted, in hybrid systems. One of the keys to the contemporary energy-saving philosophy in buildings is to put the accent on passive measures and complement the necessities with active devices.

The different climatic conditions on earth lead to different requirements in managing an optimum balance of temperature and humidity in the buildings: either a prevalence for cooling the space (in the tropical areas), for heating it (in the polar climate), or combining both cooling and heating strategies in the temperate, continental, and dry parts of the planet.

© The Author(s), under exclusive license to Springer Nature Switzerland AG 2024
A.-M. Dabija, *Architectural Design Strategies for Saving Energy in Buildings*,
https://doi.org/10.1007/978-3-031-73541-7_4

In Canada, for instance, heating as well as cooling are required, representing around 65% of the sector's energy consumption [1]. The IRENA[1] presents the same data [2]; the European Commission states that heating and cooling operations account for around 50% of the energy consumption of a building [3, 4]. The same percentage is found in the case of South Korean statistics [5]. Countries with hot and humid climate—like Malaysia—indicated a percentage of around 23% of the energy for cooling demands [6].

In all these geographic areas, indoor comfort was accomplished by less sophisticated and fuel-consuming devices. At this specific moment in history, when the emphasis is set on energy saving and on controlling the carbon footprint (as embodied energy is a part of the overall energy that defines a product), the traditional knowledge should be, once again, dusted, re-examined, and integrated at the early stages of the design process of the buildings.

Shaping the built space for maximizing the well-being of the occupants with the minimum use of resources would mean:

– Designing to reuse—in other words, taking into consideration the possibilities of assembling—disassembling, moving, and reassembling building components.
– Designing with nature—in other words using the geo-climatic characteristics of the site (and the interactions with the built environment) to establish the location, orientation, the spatial configuration of the building, and of the rooms in order to provide not only safety and protection to the users but also quality indoor conditions that provide health and well-being for the occupants: air quality, thermal comfort, visual comfort, and acoustic comfort. Among the traditional leverages that provide indoor comfort heating, ventilation and air conditioning are mandatory, and they were achieved in different cultures by specific means, depending on the geo-climatic conditions.

Both aspects—naturally provided indoor comfort as well as assembling-disassembling building components—have been invented long before our times; in some cases, probably long before humans ever existed.

Taking a step forward in the endeavor of comparing buildings with living organisms and the different types of builders, from the artisans to the trained specialists that coordinate multidisciplinary teams, we will zoom-in, on the amazing world of the non-human constructors. They too use the local resources; they, too, rely on circularity by adapting, repairing, and reusing; they, too, turn to the same technologies that have proven to be the right ones, for—probably—the entire life of their species.

Passing on the principles of construction from one generation of artisans to the other is not necessarily a human approach, as indeed humans are not the only builders of the planet: the astonishing structures carried out by "the humble world"—the buildings erected by tiny creatures—are amazing as there is no regular school or theoretical philosophy behind them. It may be the result of millenary struggles and

---

[1] The International Renewable Energy Agency.

experiments that led to some type of "rule of the thumb" that requires no more changes. They refined the skill and craft and, for many more millennia then us humans, construct their buildings over and over again, according to the species' necessities. In these cases circular economy principles as well as energy efficiency are objectives that should be reached, not as numerical applications or by using active devices—as neither did the perennial architecture of the vernacular—but to meet the necessary physical conditions of life that provide the survivance of their breed.

Keeping the proportions, if one compares some of the lodgings erected by other living beings with some of the contemporary constructions built by humans the parallel is surprising: some animals have a sense of building that cannot be explained by science; on the other hand, it is true that building their lodgings in the right way is the condition for the survival of the species. Still the question remains—and the explanation given by biologists, that "the basic building behaviour being transmitted through the genes of reproductives" is not fully acceptable (at least not by the author of this book): how DO these creatures react so logical and dynamic to the environmental conditions?! How do termites know how to build in accordance to the rules of nature and from where do they take the knowledge to do precisely what needs to be done in specific climatic conditions, in order to provide a safe and comfortable shelter that meets all the necessary requirements for carrying-out their daily activities?!

This is why this chapter begins with a puzzle: the termite mounds, a yet not fully explained natural miracle.

The Termite Mounds: Vertical Cities with Controlled Indoor Air Quality

The termites not only build multifunctional sky-scrapers but also manage to keep indoor temperature and humidity of the different "spaces" at the necessary values.

As the termites habitat is (also) in monsoon-influenced geographic areas, their survival depends on how they manage to store and preserve food for long(er) periods of time and on how they manage to raise their "children" in the maternity/ nursery. In principle, they have the knowledge to build, for each of the functions of the mound, a space with controlled temperature and good ventilation. Obviously without central heating or air conditioning equipment. But the survival of the species is depending on the know-how of the "designers" and "builders."

The building materials used by the termite-builders are "bricks" of soil and any other material they can walk on or chew, which they bond with saliva that acts like a mortar [7]. The walls of the mounds are highly porous (37–47% air, by volume and small average pore diameter), hence accomplishing diffusive transport of gases through the mound surface while, due to the small size of the pores, providing resistance to pressure-driven bulk flow across its thickness.

Judith Korb ascertains that "In accordance with other studies,[2] the nest temperature of large mounds is relatively constant, near 30 °C all year round, with a fluctuation of less than 2 °C, despite ambient temperature fluctuations of about 35 °C" [8].

The termite mounds are consistent with todays' vertical cities definition: "A vertical city is a building that includes a variety of services and amenities that "normal" cities have, such as housing, offices, healthcare, retail stores, hotels, and more. They are very tall houses that enable us to go from horizontal cities that take up large areas of land to smaller, vertical cities. These buildings will become crucial to solve the problem of overpopulation and urbanization" [9]. As far as we know (because the truth is that we don't really know much about the co-inhabitants of our planet), the termite vertical cities include all the spaces that guarantee the necessary conditions of life, from birth to death: maternity hospital, kindergarten, school, working areas, dwellings, farms, and food-courts. Moreover, the termites act as qualified designers—architects and engineers—as well as effective builders and manage to provide the necessary indoor climate conditions (heat, humidity, and ventilation management, CO_2 ratio management) in each space. They grow their own food—forging plant material in the case of some species, growing fungus in the case of other species—and make sure that the food storage is not compromised by microorganisms such as bacteria and fungi, due to moist conditions. Different species of termites build different types of mounds, and eventually they keep their house clean through a waste management system [10, 11] without a scientific theoretical base (or who knows what they are taught in their "schools"?!).

For more than a century, scientists have studied different types of termite mounds, scattered on four continents—Africa, Australia, Asia, and America—and still do not manage to fully understand how do these creatures build their vertical cities and provide the appropriate conditions of temperature and humidity in very different functional areas that have specific climatic requirements.

Compared to human buildings (and considering the scale of the "builders"), the termite mounds are higher than the highest skyscrapers humans managed to build. "If we compare the height of an average termite nest with the size of a worker termite and adjust the scale to a human being 1.80 m (5.5 feet) in height, the termite construction would be like a 960-m (3,149-foot) skyscraper—higher than every human building in the world. For example, it is five times higher than the Great Pyramid of Egypt" [12].

These structures (Fig. 4.1) fascinated, for almost one century, people and researchers who tried to understand how they work; the indoor climate seems to depend on the size, shape, and orientation of the construction but theories on how exactly the termitariums work are still on debate. The North-South orientation is supposedly providing efficient air conditioning of the nest, due to the alternate warmth on eastern and western walls that would stimulate the daily circulation of the air [11]. Having this precise alignment, the mounds erected by the *Amitermes meridionalis* (also known as "magnetic termites") were given a specific name: magnetic mounds.

---

[2] Luscher 1955, 1956, 1961; Ruelle 1964; Noirot 1970; for M. carbonarius: McComie and Dhanarajan 1990.

Fig. 4.1 Termite mounds. (Photo left: David King magnetic termite mounds, CC BY 2.0 DEED; image source: https://www.flickr.com/photos/david55king/787908916. Photo right: Bill Bradley; image source: https://commons.wikimedia.org/wiki/File:Termite-mound-Litchfield.JPG)

The first reference that we know of was made in 1897 by Robert Logan Jack and was read before the Royal Society of Queensland on September 12, 1896[3]:

> *"From the Laura to Somerset the anthills gradually increase from three feet to such a height that I have frequently been unable to touch the top with a riding whip while standing in the stirrups. The foundations and summits are strictly north and south, so much so that a man 'bushed' could steer by them, so long as he was able to distinguish north from south. The average ratio of the length of the base to the height is as four to five, and of the breadth of the base to the height as two to five. The top is sharp but serrated, culminating in a series of sharp peaks. To give stability to the structure, however, buttresses are thrown out, like those of some species of Ficus and other trees in the tropical jungles. These buttresses inter-fere with the regularity of the plan of the foundation, but not with that of the top, as they always taper to a point when they cease to be necessary as supports. [...] Why are the anthills north and south? And yet the answer seems to me simple—even ridiculously simple. The ants (or, more properly speaking, termites) build in such a direction as to secure the maximum of desiccation."* [13]

Some hypotheses were laid down during the next decades by Martin Luscher,[4] Charles Noirot,[5] F. J. Gay, and John Calaby[6] regarding the connection between the orientation of the mounds and the interior air quality—temperature and ventila-tion—but it was in 1973, when G.C. Grigg experimentally verified the statements and published them in the Australian Journal Zoology that the theory was confirmed.

---

[3] Jack, Robert Logan, Note on the "Meridional Anthills" of the Cape Yorke Peninsula [13] https://www.biodiversitylibrary.org/permissions/

[4] M. Luscher Air-conditioned termite nests, Scient. Am. 205, 1961, pages 134–45, in [11].

[5] C. Noirot, The nests of termites in Biology of Termites, Eds. K. Krishna & F.M. Weesner, Vol 2, Academic Press Inc. New York, pages 73–125, in [11].

[6] Gay, F.J. & Calaby, J.H. Termites of the Australian Region. Biology of Termites, Eds. K. Krishna & F.M. Weesner, Vol 2, Academic Press Inc. New York, 1970, pages 393–448 in [11].

Judith Korb observed that

> *"Multiple regression models of temperatures in inhabited and intact, uninhabited mounds showed that (1) mound structure alone results in the constancy of the nest temperature, (2) ambient temperature (abiotic heat production) provides a basal minimum nest temperature, but (3) that the heat produced by the metabolism of the fungi and termites (biotic heat production) is necessary to reach optimal nest temperatures of 30 °C (Korb and Linsenmair 2000a)[…] In fact, mound architecture seems to be adapted to counteract the loss of heat. Mounds in the gallery are dome-shaped, earthen structures with thick walls which have a lower surface complexity (i.e. ratio of the real surface area to the minimal possible surface) than mounds in the savannah[7] (Korb and Linsenmair 1998a). This was also confirmed by experiments: when ambient temperatures were increased in the gallery forest by cutting shading trees, mound surface complexities increased and the resulting architecture resembled that of the savannah mounds[8]"* [8].

As early as 1962, Martin Luscher defined two types of termitariums: closed mounds that work on a thermosiphon ventilation system (Fig. 4.2) and opened mounds with a chimney-like top opening (Fig. 4.3).

Fig. 4.2 The thermosiphon flow theory, adapted from Martin Luscher. (Drawing by Ana-Maria Dabija)

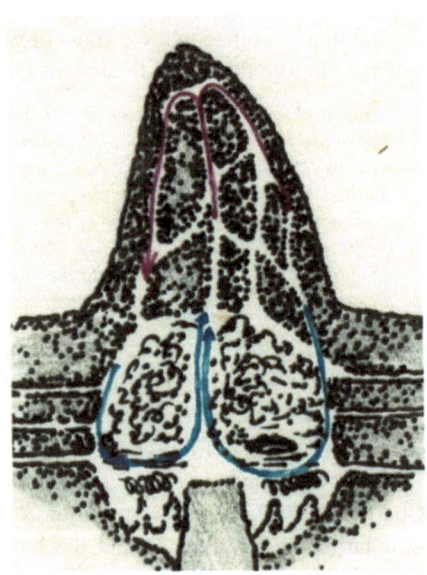

---

[7] Korb J, Linsenmair KE The effects of temperature on the architecture and distribution of Macrotermes bellicosus (Isoptera: Macrotermitinae) mounds in different habitats of a West African Guinea savanna. Insectes Soc 45 pages 51–65 in [8].

[8] ibidem.

Fig. 4.3 Cross-section of
a termite mound adapted
from J. Korb [8]. (Drawing
by Ana-Maria Dabija)

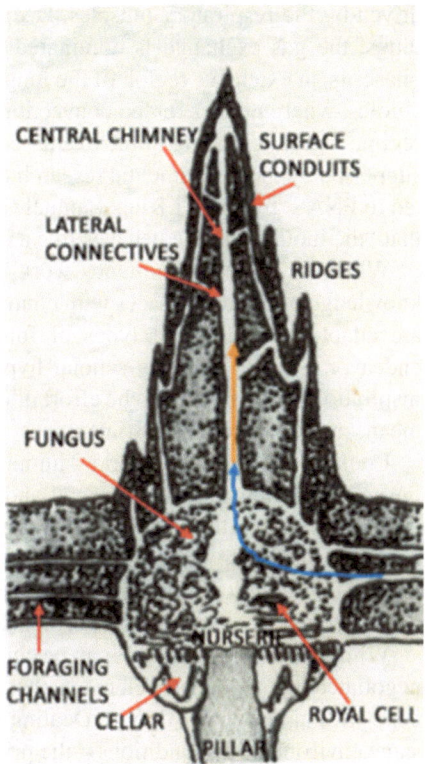

While the first typology seems easier to understand, as it works on the "stack effect," the closed mounds still puzzle the biologists who did not come to a final conclusion on how do these authentic "vertical cities" function.

Luschers' theory is that in the closed mounds, the warm air containing CO_2 rises through the channels and, reaching the upper parts just beyond the mound envelope, pushes the denser, cold, clean air down to the core of the mound. The loop closes and the cycle continues (see Fig. 4.2).

Noirot offered another explanation [11], stating that the gas changes only took place by diffusion through the walls of the building in the closed termitarium. In the same period, another theory was proposed: the wedge shape and NS axis orientation facilitates gas exchange, especially in the rainy season, as the porosity of the mound walls is reduced due to dampness. Cole launched the hypothesis according to which the gas changes occur during a convection air movement. The theory is supported in the case of mounds with "chimneys" that evacuate the warm air in the exterior. According to Hunter King and his collaborators [14], the thermal stability (and indoor air quality) is achieved due to a set of vertical channels—called "flutes"— that become warm while the central chimney is cold, thus leading to convection and gas change.

Turner and Soar compare the mounds with the human lungs that are multi-phase gas exchangers. In the first phase, gas exchange is dominated by forced convection driven by the respiratory muscles. Deeper, in the alveoli and alveolar ducts of the lungs, the gas exchange is dominated by diffusion. "Sandwiched between these phases is an extensive region of the lung, which includes the fine bronchi and bronchioles, where neither forced convection nor diffusion dominates flux. This mixed-regime region is the site of the overall control of lung function" [15]. This theory is supported by the experimental research of the species *Odontotermes obesus* submitted to PNAS[9] by Hunter King, Samuel Ocko, and L. Mahadevan [16] who conclude that "the mound surface behaves like a breathable wind-breaker."

Whatever makes the mounds work,[10] one thing is clear: the termites have the knowledge of building spaces with controlled temperature and humidity, spaces that are adapted for different types of functions, without using fossil fuels in this endeavor. Subject of many scholar hypothesis; the termite mounds have been an inspiration for architects in the effort of obtaining good indoor air quality by passive means and using local materials.

Looking at the ancient world—human or non-human—what is in fact breathtaking is that buildings were designed and constructed according to very simple and clever[11] solutions that shaped the volumetry according to the local geo-climatic restraints and using local building materials, in order to accomplish the best possible indoor conditions for the given space/function (just as the termite constructors do—keeping, of course the proportions between their world and ours).

While the anthropic agents can be considered subjective, nature's laws can not be negotiated; these are basic elements that are respected in the traditional architecture: cope with nature, don't defy it. Dealing with the same materials—and more or less same environmental conditions—the principles of composing houses were/are similar, despite the geographic location of the house. This is why sometimes buildings belonging to different cultures and situated at thousands of kilometers away from each other have sometimes a familiar aspect. Some climatic conditions unite them. Or, in other words, this is why traditional architecture seems familiar despite the distance(s) in kilometers: it uses local natural building materials that have characteristics that lead to—more or less—similar volumetric and detail solutions.

To illustrate this idea, we present examples of wooden traditional houses built in Poland, Romania, and Japan: the same climatic conditions—snow, rain, and sun— and the same building materials, wood and reed, led to the same spatial geometry of the houses and to similar construction details, although Poland is 1000 km[12] away from Romania and both of them are at about 9000 km[13] away from Japan (Fig. 4.4).

---

[9] Proceedings of the National Academy of Sciences (PNAS).

[10] "The basic building behaviour being transmitted through the genes of reproductives" according to Charles Noirot and Johanna P.E.C. Darlington in [17].

[11] We avoid using the terms "smart" or "intelligent" that are too much associated with computer applications and network.

[12] 621 miles.

[13] 5592 miles.

Fig. 4.4 Village museum, Bucharest, Romania (left). (Photo Ana-Maria Dabija. Orawski Etnographic Park, Poland (right). Photo M. Klag, CC BY-SA 2.0 DEED. Source: https://www.flickr.com/photos/mik_krakow/3534907858)

There was no theoretical approach and no international conferences, workshops, and expertise sharing between the artisans of the seventeenth/eighteenth century of Romania and Japan, yet the buildings look familiar; there was only individual experience and common sense knowledge that was passed on from one generation to another, in what we would call today "examples of good practice." The porch is a common feature, limited by the depth of the eaves that provide the protection against the summer sun while allowing the winter sun to pass through the windows. Social activities were carried out on the porch[14] and it represented an intermediate space between the private and the—more or less—public space represented by the courtyard.

The geometry of the houses, from the base to the pitched roofs, are almost identical, including the way the attic is used, as a utilitarian buffer space that requires small ventilation devices (Fig. 4.5).

Fig. 4.5 Edo-Tokyo open air architectural museum, house in Japan (left). Village museum, Bucharest, House in Romania (right). (Photo: Ana-Maria Dabija)

---

[14] Called *prispa,* in Romanian; *engawa,* in Japanese; or *ganek*—in Polish.

All the four examples have the walls raised on stone foundations, to protect the wood from rotting at the contact with the soil. The pitched roofs show that the climate includes all the four seasons and the geometry of the roof facilitates the rapid evacuation of the rain and snow; there is no difference between the wooden shingles in Poland and the wooden shingles in (some parts of) Romania. There is practically no difference between the Japanese and the Romanian reed as the thatched roofs look alike in the two houses (of the same century but located at 9000 km away).

The building materials are the same in all these regions: reed, wood, and stone.

The wooden joining details are identical, as it is also expected because in different parts of the world carpenters discovered the same best solutions of building with wood and passed them on from one generation to another. Similar geo-climatic conditions and similar materials lead to similar solutions. These systems are then replicated by generations of workers: the skill and the craft are passed from master to apprentice in years of practice. A living demonstration of how the knowledge was transmitted from one generation to another is the specific, unique case of the Grand Shrine of Ise,[15] in Japan. In other words, it is one of the very few examples of living history in the contemporary world. It also shows the preoccupation of the ancient human society to use rather than to throw away parts (of building elements, in the case of constructions) that can find other fields of application. In this case, however, both the tangible artifacts and the know-how are transmitted as a part of the cultural, philosophic, and religious tradition.

Building with the Fire

Evolving from the traditional—timeless—architecture, the principle of heating the rooms via radiant building elements, walls or floors, was discovered and considered since the Antiquity. The fire was used for cooking purposes but the hot smoke, captured, directed, and exhausted through walls or floors, warmed the rooms by radiant heating. In the continental part of Europe, the traditional houses had the kitchen stove placed close the center of the house, warming the masonry that radiated the heat to the adjacent rooms. The hot air is in this case a by-product: the main purpose of the fire was for cooking. This is why "building with fire" was framed in this chapter as a passive approach rather than an active one; as the smoke receives a distinct function before being evacuated in the exterior, the space is shaped accordingly: the location of the hearth and the plenum of the floor or walls are a resultant of the role of the hot smoke, which is to be used for better comfort of the built space (Fig. 4.6).

---

[15] More details on this topic in Chap. 3.

Fig. 4.6 Masonry stoves in traditional houses: Village Museum, Bucharest, Romania (left) Nove Hrady Czech Republic (right). (Photo: Ana-Maria Dabija (left), Jan Helebrant (right): CC BY-SA 2.0 DEED. Source: https://www.flickr.com/photos/96541566@N06/51630525552)

Ancient Korea had a similar (in principle) approach in warming the houses but instead of heating the walls, the hot smoke was directed in a "plenum" under the floor, providing radiant floor heating: it is believed to have been used since the Bronze Age (900 BC–800 BC) [18] (Figs. 4.7 and 4.8).

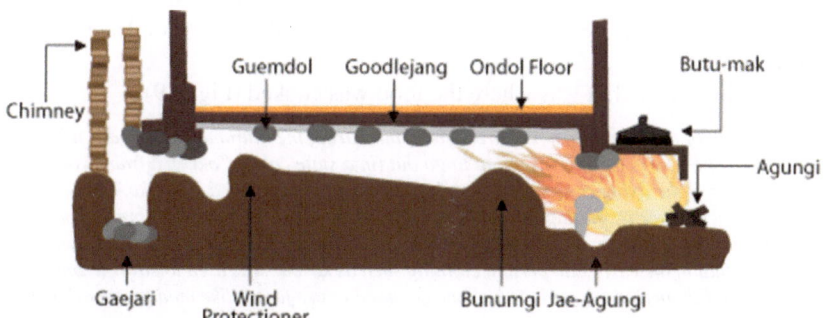

Fig. 4.7 Korean Ondol. (Photo Dzihi CC BY 3.0. Source: https://en.wikipedia.org/wiki/Ondol#/media/File:Ondol.png)

Fig. 4.8 Traditional house near Jirisan. (Photo Eimoberg, CC BY 2.0. Source: https://www.flickr.com/photos/eimoberg/214058429)

Although known—and renown—in several parts of the ancient world, the radiant flooring was abandoned in Europe after the fall of the Roman Empire but it survived in the Middle Ages, Asia, and Turkey (ottoman baths used the same principle to provide warmth in the specific rooms [19]).

Unlike the Korean system of heating the rooms by leading the hot air in a subfloor system before exhausting it through the chimney, in the Japanese[16] traditional houses, a different philosophy was applied: not to heat the air (which would also change the air humidity) but the building occupants, with a specific system [20], the *kotatsu*, which includes a heat source, a low table, and a futon. Initially, the heat source was a sunken hearth– the *irori*—located in the (more or less) center of the room.

Fueled by charcoal, it was where the food was cooked (Fig. 4.9)

"The origin of the kotatsu is not entirely clear, but during the Muromachi period,[17] it is said that when the hearth (irori) was about to go out (in a state called "oki"), a framework was built on top of the dying charcoal, covered with paper clothes called "kamiko," and people warmed themselves by putting their feet on the framework, a practice that spread from Zen Buddhism. Therefore, the framework was low, and instead of a lattice, it had a slatted top. It is said that in the early Edo period, clothing such as kosode was used to cover it, but from the middle Edo period onwards, it became common to use futons like modern times" [21].

---

[16] The Japanese traditional houses are cold in winter (except for the ones located in Hokkaido).

[17] Muromachi period (1336–1573).

Fig. 4.9 Irori (left). (Photo Shih-Chi Chiang CC BY 2.0 DEED. Source: https://www.flickr.com/photos/scchiang/34951068982. Kotatsu (right) at Rengejoin temple on Mt. Koya Photo Molly Des Jardin, CC BY 2.0 DEED. Source: https://www.flickr.com/photos/mdesjardin/2442821956)

Similar heating systems were used in Iran, under the name of *korsi* (Fig. 4.10), or Central Anatolia, Turkey, under the name of *kürsü* [22].

Fig. 4.10 Persian korsi. (Photo Maahmaah. Public domain. Source: https://en.wikipedia.org/wiki/Kotatsu#/media/File:Korsi.jpg)

Passive Design Strategies[18]

When technology was not as advanced as it is today, indoor comfort was provided with the use of passive design strategies, knowing that heating and cooling the spaces deal ultimately with the three modes of heat transfer: conduction, convection, and radiation. The envelope components with high thermal resistance and inertia were the first and obvious choice. Considering that the building envelope[19] is the most powerful "line of defense" between the interior environment and the local climatic conditions (heat, light, wind, air movement, rain, snow, earthquakes, fire, noise, etc.), most studies refer to the possibilities of improving the performances of the building envelope through passive and active measures. However, the building materials' characteristics are only the first step in the configuration of the building envelope. It is a whole system where each element has its role and contribution in accomplishing healthy indoor climate.

Whereas in the warm dominant climate pattern, the passive design measures of cooling are effective throughout the whole year, in the geographic areas of temperate climate, with a broad spectrum of air temperatures that cover more than 50 °C[20] throughout the four seasons, the set of passive means comprises both heating and cooling, therefore the supplementary active solutions (from the traditional heating of the houses, through fireplaces, heating ducts and so on, to the contemporary HVAC systems and equipment) lead to higher energy consumption needs.[21] Anthropic agents operate both inside and outside the building, increasing or diminishing the effects of the natural ones, at the scale of the building, neighborhood, and district.

Termites use—instinctively—the geographic orientation, the air movement, the sun (or shade), and the thermal mass to build and maintain their "settlements"; so did humans, especially in the traditional architecture where almost each architectural statement or gesture had to find a correspondent in the well-being of the occupants.

Therefore, the sun, the air movements, the water, and the vegetation had to be taken into consideration, in order to build houses with a better response to the climatic conditions, while keeping the living costs and the maintenance costs of the building as low as possible. We would define these, today, as "whole-life costs." Corroborated with the ingeniosity of re-using, re-shaping, and refurbishing the

---

[18] The typologies and characteristics of the materials used for the thermal insulation of the building envelope are outside the scope of this book.

[19] Defined as the building system that separates two hygro-thermal environments.

[20] As mentioned above, this is the difference of the temperature in the air; however, the measured temperature difference of the built surfaces may increase to almost or around 100 °C, on the roof surface when, during a hot summer period, the measured temperature may well rise to +70 °C ÷ +80 °C if the air temperature is around +35 °C ÷ +40 °C, while in the winter it can decrease to -15 °C ÷ -18 °C if the air temperature is around -20 °C.

[21] In countries with warm climate, the electricity demand for air conditioning may reach 23–25% of the overall energy consumption.

materials and devices—in what today we would refer to as "cradle to cradle" or "circular economy" approaches—the life conditions within the houses were surprisingly good.

Rediscovering principles that have been, if not forgotten, at least abandoned in the past century when technology based on energy seemed to be available, cheap, and endless is important as it is needed, precisely due to the goal of saving energy while maintaining the standards for indoor comfort that represent the state of well-being. This state—which has a qualitative expression—is based on the use of energy: some as a gift of nature—through passive design strategies—and the rest that cannot be achieved by passive means, complemented with the appropriate active measures. This is where the difference between passive and active means lays: passive systems rely exclusively on understanding and using the natural forces in the design of the building—and considering the biunivocal relation between the building and the environment—while active systems add technology into the nature-building equation, providing rapidly the required indoor air conditions, as long as the equipment functions.

Both passive and active systems deal with climate but in different ways: in the passive approach the built space is the direct beneficiary of the consequences of designing with nature (via its components), while in the active approach the equipment is supplied with energy from traditional or renewable resources.

It is therefore mandatory to take into consideration the same aspects and natural agents in both cases: geo-climatic and site conditions, even if they lead to different approaches.

The climatic conditions have always determined the configuration of the buildings and, according to the climatic conditions, some general "patterns" can be recognized, despite the physical distances or difference of culture. The earth has four main climatic zones, from the equator to the poles [23]: the tropical zone (between the tropics), the subtropical zone (from $23.5°-40°$), the temperate zone (from $40°-60°$), and the polar/cold zone (from $60°-90°$).

At the beginning of the twentieth century, Wladimir Köppen[22] introduced a climate classification with five colored zones, corresponding to the vegetation zones. Further developments of the Köppen[23] classification led to sub-categories, each with specific features and colors. In principle, the five climatic zones are the following:

Zone A: tropical or equatorial zone (with the temperature of the coolest month at least 18 °C).

Zone B: arid or dry zone (this zone is defined not by the temperature but by the annual precipitation falls).

---

[22] Wladimir Köppen was a Russian-born German (Austrian) climatologist and botanist who lived between 1846 and 1940; he introduced the climatologic chart of five regions, defined from A to E, mostly (but not only) on temperature criteria.

[23] According to A. Jogn Arnfield in The Encyclopaedia Britannica [24], the classification is often referred to as the Köppen-Geiger-Pohl climate classification.

Zone C: warm/mild temperate zone (with the temperature of the warmest month at
least 10 °C and the temperature of the coldest month between −3 °C and 18 °C).
Zone D: continental zone (with the temperature of warmest month at least 10 °C,
and the temperature of coldest month less than −3 °C).
Zone E: polar zone (with the temperature of warmest month of maximum +10 °C).

Highlands replicate in principle, vertically, the climate zones; more recently they
were assigned a letter for themselves: H (Fig. 4.11).[24]

Köppen-Geiger climate classification map (1980-2016)

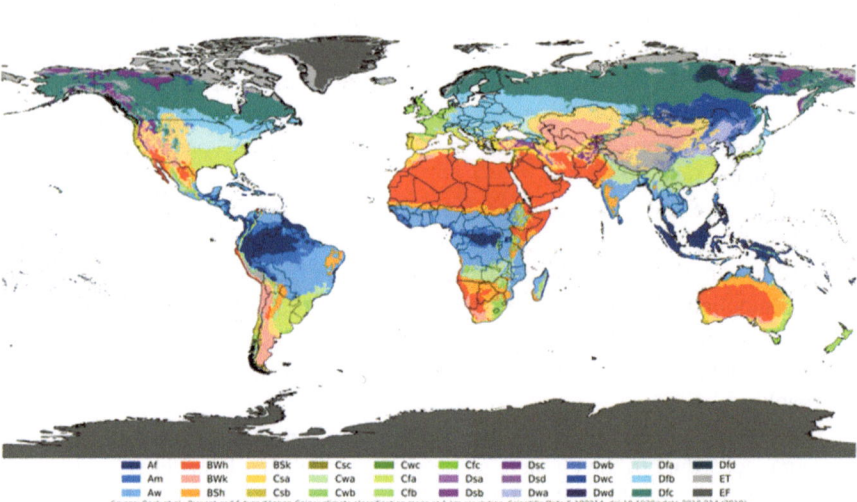

Fig. 4.11 The Köppen-Geiger climate classification (1980–2016), by Maulucioni, based on a
previous work by Beck, H.E. et al. 2018, CC BY-SA 4.0 DEED. (Source: https://en.m.wikipedia.
org/wiki/File:K%C3%B6ppen-Geiger_climate_classification_%281980-2016%29.png
Accessed 2023)

Replicating buildings on different sites is a perfect and almost infallible method
for failures: the climatic conditions corroborated with the local natural and anthropic
conditions lead to specific solutions, even though some general typologies can be
used. Adaptation to local conditions is a mandatory design phase.

Passive design strategies include intertwining principles, from the spatial configuration of building in accordance with the geo-climatic reality of the site, to the
building materials' characteristics: in the cold and continental temperate climates,
buildings are compact, the walls and roof assemblies are thick, the windows are
small in order to minimize the heat loss, the roofs are pitched to evacuate rainwater

---

[24] G.T. Trewartha, An Introduction to Climate, fourth ed. (1968), in The Encyclopaedia
Britannica [24].

and snow as fast as possible. Main spaces are oriented toward the sun for the bonus heat provided by solar radiation. The hearth is always placed in the heart of the house, radiating heat in the surrounding spaces.

In temperate climates, the porches and the eaves are dimensioned to provide the necessary summer shading, while allowing the sun-beams to enter and warm the rooms, in the winter season. Shutters add protection against the heat of the summer sun but also thermal insulation in the winter season (Fig. 4.12).

Fig. 4.12 Traditional house in Dobrogea, Romania (Village Museum). (Photo: Ana-Maria Dabija)

From the mild-temperate to the tropical regions of the world, houses are often organized around interior courtyards. The layout of the ancient Greek house included the prostas; the single-family building typology in ancient Rome included a roofless courtyard: the atrium. Usually atriums were equipped with a fountain or a basin where rainwater was collected; sometimes, in the villas of larger dimensions, the more intimate spaces of the house were located beyond the atrium, around an inner, open, courtyard that was surrounded by a colonnade, providing solar control for the adjacent rooms.

Similar layouts—but using different building materials—can be identified in traditional houses in India, also organized around inner courtyards from where light and ventilation were provided for the adjoining rooms (Fig. 4.13).

Fig. 4.13 Atrium in the House of the Wounded Bear, Pompeii, Italy. (Photo: Gary Todd Public Domain. Source: https://www.flickr.com/photos/101561334@N08/48442024151)

In the hot, dry climate, the necessity to evacuate the warmth of the house led to domed structures: the high space allowed the warm air to rise and be exhausted through apertures under the lantern.

In a nutshell, the buildings in the cold and temperate climate used passive means both for cooling the space, in summer and for heating it, in winter; the buildings in the warm climates needed to keep the interior space cool.

Building with the Sun

One of the most famous architects of the twentieth century, Louis Kahn, once said "The Sun does not realise how wonderful it is until after a room is made." In just a few words everything has been said.

As much as it may surprise us, the contribution of the sun in the well-being of the building occupants had been stated more than two thousand years ago, by the roman architect and military engineer Marcus Vitruvius Pollo in his famous treatise *De Architectura Libri X.*[25] Vitruvius's text is beautiful as it is concise and correct. And—of course—with a flavor of a civilization long gone:

---

[25] Ten Books on Architecture.

1. *"These are properly designed, when due regard is had to the country and climate in which they are erected. For the method of building which is suited to Egypt would be very improper in Spain, and that in use in Pontus would be absurd at Rome: so in other parts of the world a style suitable to one climate, would be very unsuitable to another: for one part of the world is under the sun's course, another is distant from it, and another, between the two, is temperate. Since, therefore, from the position of the heaven in respect of the earth, from the inclination of the zodiac and from the sun's course, the earth varies in temperature in different parts, so the form of buildings must be varied according to the temperature of the place, and the various aspects of the heavens.*

2. *In the north, buildings should be arched, enclosed as much as possible, and not exposed, and it seems proper that they should face the warmer aspects. Those under the sun's course in southern countries where the heat is oppressive, should be exposed and turned towards the north and east. Thus the injury which nature would effect, is evaded by means of art. So, in other parts, due allowance is to be made, having regard to their position, in respect of the heavens.*

3. *This, however, is determined by consideration of the nature of the place..."* [25].

Building orientation is a nuanced approach: there are spaces that need to be facing the sun and—*au contraire*—there are spaces that need constant lighting conditions, provided by orienting those windows toward the Poles (north in the northern hemisphere, south in the southern hemisphere). By simple orientation, efficient use of energy resources can be accomplished, as—for instance—no need for solar protection and artificial light would be required.

Moreover, there are spaces that need to have constant, and low(er) temperature, like the storage rooms (for instance): the appropriate orientation—facing the poles—would reduce the need of cooling them with active means (refrigeration, air conditioning, and similar).

The equator-facing orientation (south for the northern hemisphere and north for the southern hemisphere) on the other hand is providing the best solution for passive solar gain: as the light strikes terrestrial objects, the wavelength changes; objects warm up as they absorb heat and they eventually re-radiate it at longer (infrared) wavelength.

It is an observation that led to the orientation of the main rooms toward the light, since the Antiquity, and was extensively explained by Vitruvius who continues, in this chapter of the same book [25]:

1. *"I shall now describe how the different sorts of buildings are placed as regards their aspects. Winter triclinia and baths are to face the winter west, because the afternoon light is wanted in them; and not less so because the setting sun casts its rays upon them, and but its heat warms the aspect towards the evening hours. Bed chambers and libraries should be towards the east, for their purposes require the morning light: in libraries the books are in this aspect preserved from decay; those that are towards the south and west are injured by the worm and by the damp, which the moist winds generate and nourish, and spreading the damp, make the books mouldy.*

2. *Spring and autumn triclinia should be towards the east, for then, if the windows be closed till the sun has passed the meridian, they are cool at the time they are wanted for use. Summer triclinia should be towards the north, because that aspect, unlike others, is not heated during the summer solstice, but, on account of being turned away from the course of the sun, is always cool, and affords health and refreshment. Pinacothecæ should have the same aspect, as well as rooms for embroidering and painting, that the colours used therein, by the equability of the light, may preserve their brilliancy."*

Vitruvius treats the town residences different from the country ones, where people and animals must be accommodated in appropriate shelters.

Therefore he analyzes, in Chap. 6 [25], the house and household from the perspective of the salubrity, functionality, as well as comfort and maintenance, putting emphasis on *light*[26]:

1. *"First of all the salubrity of the situation must be examined, according to the rules given in the first book for the position of a city, and the site may be then determined. Their size should be dependent on the extent of the land attached to them, and its produce. The courts and their dimensions will be determined by the number of cattle, and the yokes of oxen employed. The kitchen is to be placed in the warmest part of the court; adjoining to this are placed the stalls for oxen, with the mangers at the same time towards the fire and towards the east, for oxen with their faces to the light and fire do not become rough-coated. Hence it is that husbandmen, who are altogether ignorant of the nature of aspects, think that oxen should look towards no other region than that of the east.*

2. *The baths should be contiguous to the kitchen, for they will be then serviceable also for agricultural purposes. The press-room should also be near the kitchen, for the convenience of expressing the oil from the olive; and near that the cellar, lighted from the north, for if it have any opening through which the heat of the sun can penetrate, the wine affected by the heat becomes vapid.*

3. *The oil room is to be lighted from the southern and warmer parts of the heaven, that the oil may not be congealed, but be preserved liquid by means of a gentle heat.*

4. *Care should be taken that all buildings are well lighted: in those of the country this point is easily accomplished, because the wall of a neighbour is not likely to interfere with the light. But in the city the height of party walls, or the narrowness of the situation may obscure the light. In this case we should proceed as follows. In that direction from which the light is to be received, let a line be drawn from the top of the obstructing wall, to that part where the light is to be introduced, and if, looking upwards along that line, a large space of open sky be seen, the light may be obtained from that quarter without fear of obstruction thereof;*

5. *but if there be any impediment from beams, lintels, or floors, upper lights must be opened, and the light thus introduced. In short, it may be taken as a general rule, that where the sky is seen, in such part apertures are to be left for windows, so that the building may be light. Necessary as light may be in triclinia and other apartments, not less is it so in passages, ascents, and staircases, in which persons carrying loads frequently meet each other."*

From the vernacular houses located in the cold climates all around the world to the luxurious baths in the villas of the Roman empire, designing with the sun and using its potential to warm the rooms was a natural step to take. In fact, the target was warming people not warming spaces. And this is an interesting approach: life oriented and therefore efficient in terms of the use of resources. And even this idea only, launched by Vitruvius, could bring a new perspective on the contemporary building philosophies.

With or without knowing Virtuvius's writings, architects—known or unknown artisans—respected these principles until the consumerist twentieth-century period (when the motto was, more or less, "man is the master of nature"). Today they should be considered one of the design leverages of avoiding the depleting of natural resources (and energy is one of them).

---

[26] In fact, the noun is used no less than 17 times throughout the 54 paragraphs of the 8 Chapters: 3 times in Chap. 4 and 14 times in Chap. 6.

Building with the Sun: Enhancing the Light

One of the most famous buildings of the ancient Roman architecture is the Terme within the Villa Adriana in Tivoly, dating from the first-century AD.

Designed according to Vitruvius's theories exposed in Book V[27]"1. First, as warm a spot as possible is to be selected, that is to say, one sheltered from the north and north-east. The hot and tepid baths are to receive their light from the winter west; but, if the nature of the place prevent that, at all events from the south, because the hours of bathing are principally from noon to evening" [25], the Roman Baths in Tivoli increase the intake of warmth as the south and east openings were equipped with glass panes [27]. As previously mentioned, once the sun beam passes through the glass and strikes material surfaces, it changes the wavelength and warms the space, by re-radiation of the heat of the solid elements (Fig. 4.14).

Fig. 4.14 Heliocaminus in Villa Adriana in Tivoli, first-century AD. (Photo: Claire Cox (CC BY-ND 2.0) source: https://www.flickr.com/photos/clairemcox/47985484957)

---

[27] On the different types of public spaces and buildings: The Forum and Basilica, The Treasury, Prison and Senate House, The Theatre, Colonnades and Walks, Baths, The Palaestra, Harbours, Breakwaters and Shipyards [26].

Directing the light into dark places required knowledge and skill: light shelves are not a very old invention compared to the timeless eaves, because they rely on the flatness, smoothness, and color of the material that acts as an element that, by the controlled multiple reflections of the light, pushes the sun-beam into the depth of the building.

However, before the light-shelves were invented, controlled light reflections were accomplished by mirrors or mirror-like metallic shields. And they can be tracked down to the Greek Antiquity.[28] One hundred and fifty years ago, a huge leap was taken by Paul Emile Chapuis, a photographer who, under the slogan "Why burn gas?" sold his patented Daylight Reflector.[29] As can be seen, the problem of using gas is not new; only that two centuries ago it had to do with the costs that the owners/tenants had to pay to run a building while today the perspective moves from an individual scale to a world management scale.

The use of oriented, flat, highly reflective surfaces rediscovered by Chapuis led to an entire world of products and systems: light shelves (Fig. 4.15), light guiding shades (Fig. 4.16), anidolic systems (Fig. 4.17), light pipes, etc. All of them with the same goal: to bring light into dark places and to reduce the energy bill that would have been paid by the owners/administrators/tenants of the building(s).

Fig. 4.15 Light shelves. (Dabija, A-M., Alternative Envelope Components for Energy-Efficient Buildings Chap. 4: The Sun—Building Partner of All Times; Passive and Active Approaches [29], page 70)

---

[28] The legend of how the Greeks dismissed the Roman attack during the siege of Syracuse in 213–212 BC thanks to Archimedes' "burning mirrors" is widely known but is also worth remembering; see [28].

[29] [29], page 68.

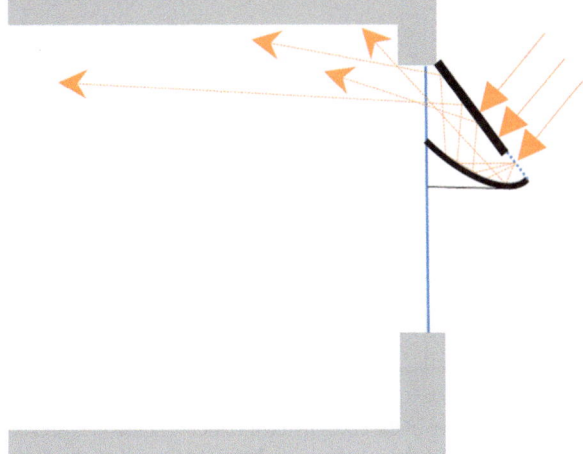

Fig. 4.16 Light guiding shades. (ibidem [29] page 72)

Fig. 4.17 Anidolic ceiling. (Ibidem [29] page 72): *1.* roller blind; *2.* double glazing; *3.* light concentrator; *4.* duct; *5.* ceiling; *6.* transparent (ceiling) element (daylight distribution element); *7.* interior light concentrator

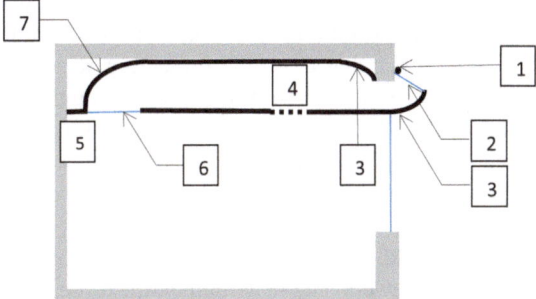

Building with the Sun: Filtering the Light

While the sun was a "building means" to contribute to the comfort of the living beings—human as well as non-human—it is, in warm climates, a factor of discomfort that should be avoided or filtered by different types of sunshades: blinds, eaves and—when and where appropriate—sunscreen panels.

The sunshades are—as their name suggests—elements that protect the interior space against the aggressive sun-beams of the summer. However, the winter sun is more than welcome in the interior of the building. Therefore they are dimensioned taking into consideration both the solar angle in the winter and in the summer. The result is an architectural element that allows the sun to penetrate the glass in the winter and shades it during the summer (preventing the overheating of the interior space): the eave (Fig. 4.18).

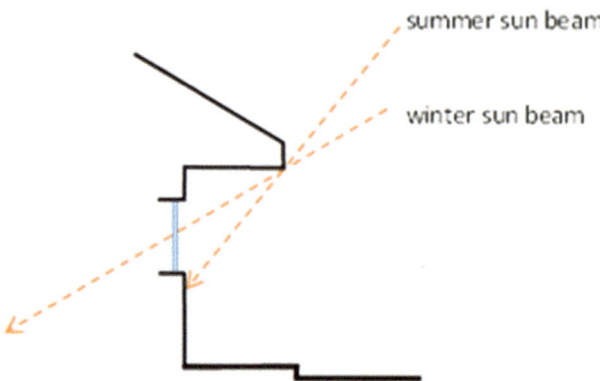

Fig. 4.18 Schematic presentation of the eaves in the traditional house: it is dimensioned to provide protection of the window against the summer sun beam and to allow the winter beam into the room, through the window. (Ibidem [29] page 64)

The sunscreens have the role of filtering the light. They are specific protections of the openings in the exterior walls, in hot and dry climates. The traditional architecture of the Middle East abounds in moucharabiehs or mashrabiya—spectacular wooden laces[30] that close the projecting windows (bay windows) and filter light while allowing the air to move freely through the interior spaces of the building, thus regulating the temperature and humidity of the building. In relation to the architectural space, these lattice screens also provide a remarkable decorativeness both inside and outside the building and ensure the privacy of the interior space over the exterior.[31] Putting it in very simple words, the mashrabiya is a window that connects the building occupants with the outside world (but not the outside world with the indoor viewers) and in the same time represents a protection against the dazzling sun of the hot climate area where the building is constructed. The apertures that form the screen are designed to allow the air to flow into the room and in the right direction, providing the chamber(s) with a natural air conditioning system (free of any energy consumption costs). A European equivalent—with less decorative features and—would be the louvers that protect the interior space against the sun.

Returning to the mashrabiyas, similar devices can be found in the traditional architecture from the Far East to South America: India, Japan, China, Portugal, and Spain [30].

It seems [30] that the oldest mashrabyia that survived the challenges of time is the Minbar[32] of the Great Mosque of Kairouan (Fig. 4.19), a city in Tunisia. It dates from the thirteenth-century AD.[33]

---

[30] Stone, brick, and plaster have also been used for mashrabiya but less frequently.

[31] Eventually they evolved as interior screens as well, separating different types of spaces.

[32] The pulpit from where the imam delivers his sermon.

[33] Other authors place it in the eight-century [33].

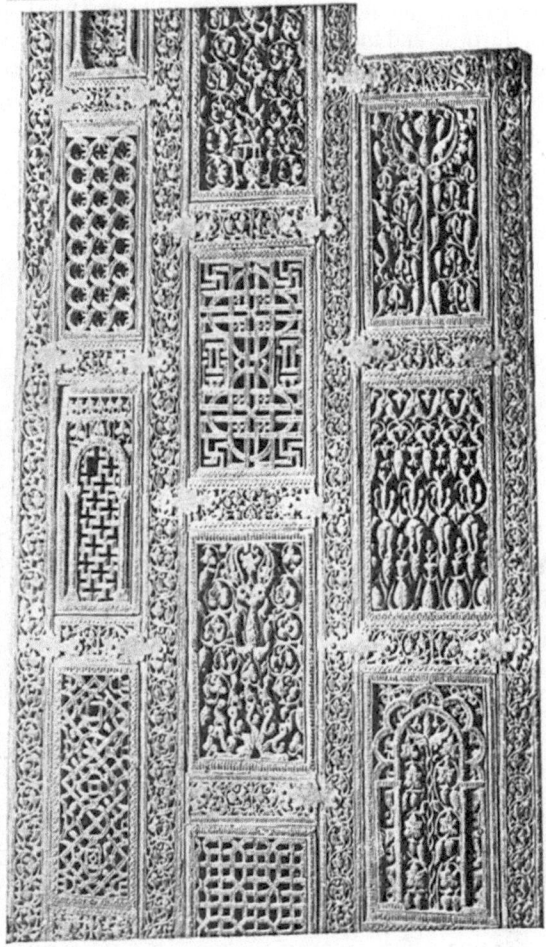

This spectacular wooden lace was created as a functional/utilitarian element, evolved as one of the most characteristic features of the Islamic architecture and from there, transcends the physical, geographic borders and replicates into the modern and contemporary architecture. According to Taşkan and Behzad Ismaeel [31] *"mashrabiya is called "mashrabiya" in Egypt, Syria, Palestine, Lebanon, Sudan, Australia, and Peru; "roshan" in Gulf Arabic Countries and Iran; "şanşol" or "şanaşil" in Iraq; "jali" in India, Pakistan, and Bhutan; "aggasi" in Bahrein; "takrima" in Yemen; "mushabek" in Iran; "barmaklı" in Maghreb countries and "cumba" in Turkey."* "The linguistic origin of the term roshan is Farsi, and it means a source of light. The historical origin is based upon the word roshān from the very distant past" [32]. Al-Murahhem also analyzes the specific structure of the wooden building lattice: "This wooden garment that wraps the skin of the building has unique parts and is constructed in a way that mimics the features of the human face.

That is, the upper part could be seen as the forehead, the middle part mimics the eye and the nose and the lower part as a chin. Each part, with its ornamentation, fulfils a specific function from within the house and keeps it cold and airy" [32, page 561]. "Structurally, the roshān is usually completed with three parts (the lower part, the middle part where the openings are located and the upper part); each part has its local name. It could be seen as one separate unit to cover an opening or a side of a room, or a continuous one that envelopes the whole façade of the house" [32, page 562].

The light and air control is provided by the moving elements of the middle part of the mashrabiya (Fig. 4.20).

Fig. 4.20 Mashrabiya in Old Cairo. (Photo: Sam Valadi CC BY 2.0 DEED. Source: https://www.flickr.com/photos/132084522@N05/16952841285)

The laced screen is a part of a building system where each component has its precise role in meeting the necessary parameters of air and light control:

– Light is filtered by the perforated screen, and glare is avoided.
– Indoor temperature and humidity are controlled[34] due to the cumulated effect of the building configuration, thick walls, and screens. The walls supply the thermal mass, absorbing heat during the day and releasing it at night. The wooden screen allows the passage of the air, by refined orientation, proportions, and shapes of the wooden elements, while filtering the dust. The geometry of the building which in the Middle East includes vaulted spaces and inner courtyard directs the

---

[34] Although diurnal temperature variations can reach 40 °C in hot, arid climates.

hot air upward and evacuates it through the wind towers, permanently "washing" the interior with a fresh supply of clean(er) air. Even the characteristics of the building material has a contribution to the regulation of the indoor air humidity: "It is known that wood absorbs, retains and releases water. When air passes through the interstices of the porous wooden mashrabiya, it vaporises some of the moisture gathered in the wood and carries it towards the interior"[30, page 7]. Good performance for ventilation and dust filtering is achieved by tilting the louvers of the mashrabiya at specific angles, according to the local conditions [30, page 7].

Building with the Wind: Ventilation, a Mandatory Condition for Well-Being

Chapter 2 approached briefly what is known in the scientific world as "the sick building syndrome"; in other words, what happens if a built space is not properly ventilated. The sick building syndrome is related to buildings equipped with mechanical ventilation or air conditioning systems. While the history and evolution of the HVAC goes back around three centuries (and air conditioning is one century old), the houses of the hot arid—or humid—climates, for instance, were cooled down long before the invention of the active systems. According to El-Borombaly and Molina-Prieto "Ventilation systems and natural cooling are as old as architecture, and are present in human settlements in warm climates around the world. [...] In Egyptian papyrus 3500 years ago these systems are represented, in Iran its use dates back 4000 years BC, in Babylon since 600 BC, and in Sri Lanka a similar concept in utilized in the primitive tents. But in other side of the world, also they used since ancient times. The Amazonian communities of South America used wind energy to ventilate their Longhouse more than 5000 years ago, and the Indians of Mochica in Peru using the natural systems by ventilate and cool their homes (Brañas, 2015; Komijani et al, 2014; Pirhayati et al, 2013; Bahramzadeh et al, 2013)" [34].

Where climatic conditions permit, natural ventilation may be a better option than mechanical ventilation, both in terms of efficiency and energy costs: it is free (including no maintenance costs), and the risk of airborne contagion is low (as seen before, the sick building syndrome is related to the mechanical ventilation and air conditioning equipment).

Natural ventilation can be accomplished by the following systems that can be used alone or combined:

– Single-side ventilation (Fig. 4.21), by means of operable windows. In other words, open the window and ventilate the room. However, the performance of the system depends on the users (if nobody opens the windows, no ventilation occurs).

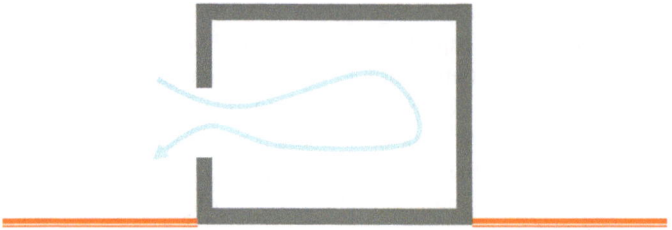

Fig. 4.21 Single-side ventilation

– Cross-ventilation (Fig. 4.22), by positioning the inlet and exhaust apertures (windows, grids, and louvers) to maximize the air passage; the air flow must enter the building, cross the room(s) and exit on the opposite side, eliminating the viciated air as well as the pollutants. Well-designed[35] and installed, cross-ventilation systems provide a high ventilation rate and are energy efficient but only in the appropriate climatic conditions. One of the drawbacks is that, like in the case of the single-side ventilation, someone (including a computer software) has to operate the windows/apertures, at the appropriate or established time; as the different spaces must communicate through the apertures, in the case of cross-ventilation care should be taken in respect with the acoustic insulation: each opening—door, louver, and grid—would transmit the sound from one space to another.

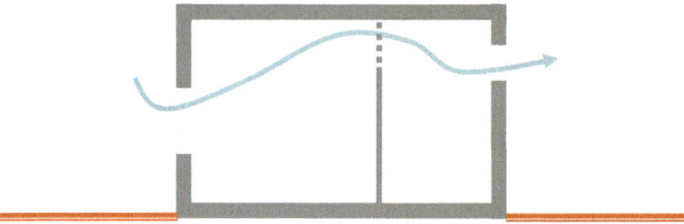

Fig. 4.22 Cross-ventilation

– Stack (or buoyancy) ventilation, based on temperature differences that move the air upwards (Fig. 4.23).

---

[35] Computer modeling and simulations can lead to a more precise positioning of the operable elements, maximizing the effect of the ventilation system.

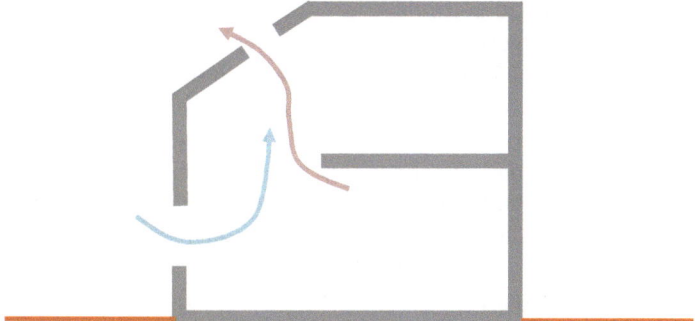

Fig. 4.23 Cross-ventilation

Cross-ventilation is the capacity of moving the masses of air horizontally, while stack ventilation is the capacity to move the air masses vertically.

Both systems aim to introduce fresh air from the exterior and remove stale indoor air but the stack ventilation accomplishes one extra function: to remove warm(er) air from the building.

The ability to control the air movements was one of the tools of the architects—known or unknown—to provide, through design, healthy indoor air conditions. A common denominator of these solutions is the chimney.

Chimneys

The chimney is such a trivial building element that nobody notices it any more. It is a building component with the role of evacuating the hot gases and smoke produced in winter by a functioning fireplace; however, it has a similar role during the summer, as it is the path the foul air is evacuated from the rooms. Analyzing the chimney from a more scientific perspective, it results that its functioning is based on the difference of temperature and the difference of pressure of the air between the low and the upper part. In other words it cumulates the stack effect[36] and the Bernoulli principle.[37] While the stack effect is based on the difference of temperature between the interior and the exterior, the Bernoulli principle is related to the air movement and

---

[36] Defined, according to [35], as "the movement of air in and out of buildings due to differences in the temperature or moisture content of air".

[37] It derives from the principle of conservation of energy [36]. As in a steady flow "the sum of all forms of energy in a fluid will be the same at all points of that streamline, while all the energy remains constant, an increase in the speed of the fluid will imply there is an increase in the dynamic pressure (kinetic energy). This happens with a simultaneous decrease in the potential energy including the static pressure and internal energy. States that increasing the velocity of a fluid leads to decreasing the pressure of the fluid."

wind around the construction. In other words, both the natural and the anthropic environment have an influence on how the air moves within a building.

Chimneys are associated with cold climates, as their main, obvious function is to evacuate the hot smoke and gases that are produced by combustion. Hence, the role of air circulation, fulfilled in all the other moments of the day, is often neglected: warm air is less dense and therefore rises and is eliminated in the atmosphere through the chimney. In other words, chimneys have an essential role in the winter as well as in the summer, to provide good indoor air quality.

From a functional device, architects transformed the chimney into an aesthetic one, with architectural value integrated in the volumetric composition of a building (Fig. 4.24). One "classic" example is the Casa Batllo,[38] in Barcelona, designed by architect Antoni Gaudi in the early twentieth century: the chimneys have sculptural shapes and are cladded with ceramic tiles, participating at the architectural expression of the building (Fig. 4.25).

Fig. 4.24 Chimneys. (Photo: Clive Varley CC BY 2.0 DEED. Source: https://www.flickr.com/photos/boyfrom_bare/26876782599)

---

[38] More about this building in Chap. 5

Fig. 4.25 Arch. Antoni Gaudi: Casa Batllo, Barcelona, 1904. (Photo: Henry Burrows CC BY-SA 2.0 DEED. Source: https://www.flickr.com/photos/foilman/48490767097)

In order to provide good indoor air, both conditions should be met: difference of the air temperature and of the air pressure, in other words, windows (or other openings) as well as chimneys, or towers. In fact, the towers, the atriums are chimneys at a very much larger scale.

The stack effect has been used in architecture as a means of pulling warm air out of the building, extending the role of the chimney from a local element of exhausting hot gases to an architectural element that provide a better indoor climate without using active devices: inside the building, as it gets warm, the air also gets loaded with pollutants (from human activities, from the building materials, or equipment) that will also be evacuated through the chimney-like building elements where the stack effect functions. The Bernoulli principle applies where wind is (also) involved, the moving, exterior air managing to suck the air with higher pressure from the building.

Wind Towers

The wind towers (or windcatchers, as they are commonly referred to) are large-scale "chimneys" with the function of providing natural air conditioning in hot climates. They are an important feature of the traditional housing systems located mainly but not exclusively in the Middle East: similar conditions lead to similar solutions, despite the physical distance; therefore related constructions can be found, bearing different names but accomplishing the same functions, from Egypt and the Middle East to Pakistan and India. In the Persian Gulf area, the windcatchers are called *badgir* while in Egypt and Iran they are called *malqaf* [37]. In Pakistan the wind-catcher is reffered to as *manghu* [38].

It seems that they have been first used in Iran [39].

As previously mentioned, these chimneys are a part of the construction system of the building. The building materials and the thickness of the walls are important features, as they provide thermal inertia and delay the transmission of heat from the environment to the inner space by several hours (up to 12 hours, in the case of 35–40 cm thick brick walls) (Fig. 4.26).

Fig. 4.26 Wind towers in Yazd, Iran. (Photo: Ninara, CC BY-SA 2.0 DEED. Source: https://www.flickr.com/photos/ninara/8666742074)

Wind towers function according to the same principles as the chimneys, based on the difference of the density of the air (related to temperature and humidity) and on wind speed. The greater the thermal difference and the height, the greater the stack effect. Warm air is lighter and rises, reducing the air pressure at the base level of the building and allowing fresh air inside. There is a difference in the average height of the windcatchers, according to the typology of the climate: in hot, dry areas the "chimneys" are higher than those of the hot and humid geographic areas, as the greater the difference of air pressure, the better buoyancy is accomplished.

In order to lower the indoor temperature and to increase the air humidity, the windcatcher is often constructed above a water supply (Fig. 4.27): a reservoir, a *qanat* (system of shafts and tunnels that reach an underground water source and transports it by gravity to a location where it is used for agriculture and domestic use), a pond, or a *salasabil* (fountain connected to a marble, wavy stone, hence dissipating the water on a larger surface) [37].

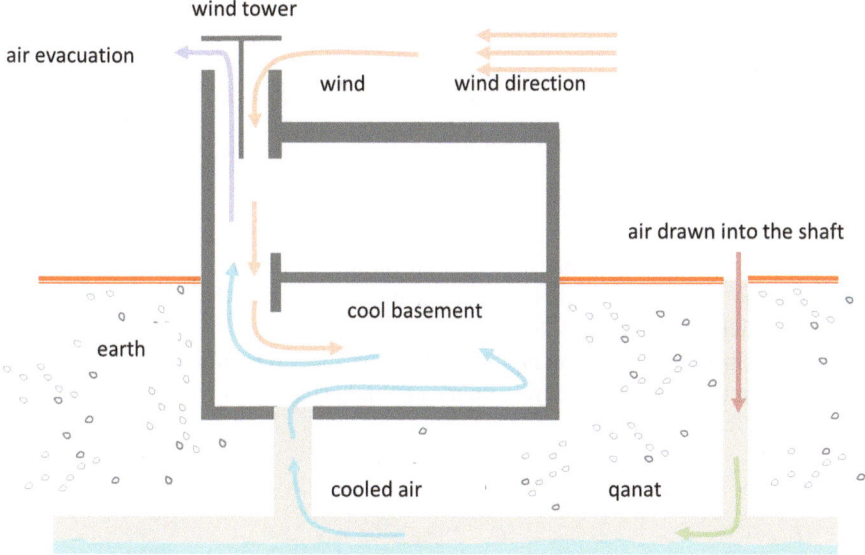

Fig. 4.27 Bidirectional wind tower

Unlike the regular chimneys that begin above the hearth, the base of the wind tower is in the basement (Fig. 4.27). Like the chimney, the wind tower rises over the roof with a specific height that takes into account the local environmental conditions (and the influence of the anthropic input). While the role of the chimney is to exhausts hot and toxic gases produced by the fire, the role of the wind tower is to circulate "coolth" provided by the underground water.

Some researches [37] consider that both *badgir* and *malqaf* are bidirectional systems, with the difference that the *badgir* functions only in summer and is closed during the winter [37] while the *malqaf* is combined with a *salasabil*, a marble plate linked to a water source.

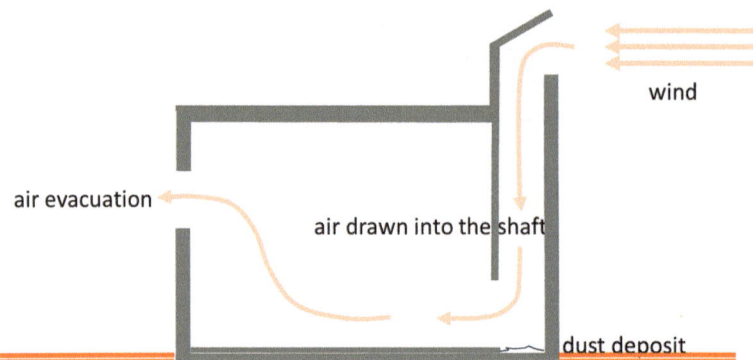

Fig. 4.28 Malqaf

According to other sources [40], the *malqaf* is a shaft that rises above the building with a lateral opening that faces the prevailing wind (Fig. 4.28), catching and directing the air toward the lower areas of the shaft, while the *badgir* is a multidirectional windscoop (Fig. 4.20).

As seen in Fig. 4.29, the *manghu* (*malqaf*) was a common urban feature in the late 1800, in the warm and humid climate of Sindh (Pakistan).

Fig. 4.29 Hyderabad, Sindh, late 1800s. (Photo: public domain. Source: https://en.wikipedia.org/wiki/File:Hyderabad1800s.jpg)

Its position within the functional scheme of the building is at the entrance (logical, as it has to improve the indoor air composition for the users as they enter the house) but there are houses where each room is provided with a windcatcher.

The height of the windcatcher is a matter of representation as well as of functionality: richer families have houses with higher towers that function better due to the difference of temperature and air pressure. Besides, as the apertures are higher, the air is cleaner (carries less dust). The highest windcatcher [39] is 33 meters high and is located in Yazd in Dowlat Abad Garden (Fig. 4.30).

Fig. 4.30 Badgir (multidirectional windcatcher) in Dolat Abad Garden, Yazd, Iran. (Photo: slingshot, CC BY 2.0 DEED. Source: https://www.flickr.com/photos/s1ingshot/21087329512)

The apertures of the *badgir* should be facing south and north: the sun warms the southern side of the tower, hence it eventually warms the air inside (due to conduction in the masonry and convection between the solid—the wall—and the fluid, the air) that becomes lighter and rises; its place is taken by cooler air coming from other interior spaces and from the openings facing the northern side of the tower.

At night, as the temperature falls significantly, the tower cools and the air temperature in its vicinity is higher than the ambient temperature; the warm(er) air rises and exits through the openings. The process continues until the temperature of the wall and the temperature of the air outside the building become close to equal (phenomenon which occurs at the end of the night).

Building with Water

Water and vegetation have often been leverages to improve the living conditions of the built space, not only inside the buildings but also at the urban level.

The management of water led, since the Antiquity,[39] to the development of roof-top gardens, parks, and oases.

Inside the buildings, the fountain and the salasabil[40] increase the humidity of the space and contribute to achieving indoor air comfort through evaporative cooling. Unlike the fountain that is adopted where water is abundant, the *salsabil* maximizes the water surface, where the water source is thin (Fig. 4.31): it is a decorated stone with a sloped surface on which the water spreads; ventilation maximizes the evaporation and cools the air.

Fig. 4.31 Salsabil in the House of the Large Fountain House, Pompeii. (Photo: Mary Harrsch CC BY-SA 2.0 DEED. Source: https://www.flickr.com/photos/mharrsch/51507783742)

The hot water of the geothermal springs has been used for more than two thousand years as a means to provide warm water and radiant floors: ancient Pompeii channeled the resulting hot steam—where available—into the plenum of the floors, to heat the spaces [41].

---

[39] [29], page 32

[40] Also known as *salsabil*

France is the country that documented the installation of the world's first district heating system, in Chaudes-Aigues, in 1332 [42–44]. Even the name of the village means "Hot Water," as there are 30 hot springs, with temperatures ranging from 45 °C to 82 °C (spring Par).

In the fourteenth century, the locals created a system of wooden pipes that collected and directed the hot water under the stone floors. Apparently, the system is still functioning.

More or less similar solutions have been practiced in other locations in the world, where the natural hot water source existed. Today the geothermal energy is used in active systems of producing clean energy.

References

1. Heating and Cooling Strategies in the Clean Energy Transition. Outlooks and lessons from Canada's provinces and territories. Published May 2019 https://www.iea.org/reports/heating-and-cooling-strategies-in-the-clean-energy-transition. Accessed 2023.
2. Power to heat and cooling: Status in https://www.irena.org/Innovation-landscape-for-smart-electrification/Power-to-heat-and-cooling/Status. Accessed Dec. 2023.
3. Energy, Climate change, Environment, official website of the European Commission, https://energy.ec.europa.eu/topics/energy-efficiency/heating-and-cooling_en. Accessed Dec 2023.
4. Heating and cooling from renewables gradually increasing, Eurostat, 3 Feb. 2023, https://ec.europa.eu/eurostat/web/products-eurostat-news/w/DDN-20230203-1. Accessed Dec 2023.
5. Kim, D. D., & Suh, H. S. (2021, April). Heating and cooling energy consumption prediction model for high-rise apartment buildings considering design parametersEnergy for Sustainable Development, 61, pp. 1–14, https://doi.org/10.1016/j.esd.2021.01.001, Accessed Dec 2023.
6. M Al-Obaidi, K., Ismail, M., & AbdulRahman, A. M. (2014, September). Passive cooling techniques through reflective and radiative roofs in tropical houses in Southeast Asia: A literature review, Frontiers of Architectural Research, 3, 3, pp. 283–297, https://doi.org/10.1016/j.foar.2014.06.002 Accessed 2022.
7. Pendharkar, V. (2017). A method in the madness: How termites build and repair their mounds. The Wire. Available at: https://thewire.in/science/termites-mound-bolus-granular-hydrogel. Accessed 2022.
8. Korb, J., (2003, February 11). Thermoregulation and ventilation of termite mounds, Die Naturwissenschaften, 90(5), pp. 212–219, https://doi.org/10.1007/s00114-002-0401-4. Accessed 2022.
9. Rosenkranz, E. (2021, August 1). What is a vertical city?—Definition and features in https://smart-cre.com/what-is-a-vertical-city-definition-and-features/. Accessed 2022.
10. Schmidt, A. M., Jacklyn, P., & Korb, J. (2013). 'Magnetic' termite mounds: Is their unique shape an adaptation to facilitate gas exchange and improve food storage?, Insectes Sociaux International Journal for the Study of Social Arthropods, International Union for the Study of Social Insects (IUSSI). Published online October 2013, https://doi.org/10.1007/s00040-013-0322-6. Accessed 2021.
11. Grigg, G. C. (1973). Some consequences of the shape and orientation of "magnetic" termite mounds. Australian Journal Zoology, 21, 231–237.
12. Hoyt, E., & Schultz, P. (1999). Insect life (p. 160). Wiley.
13. Jack, R. L. (1897). Note on the "Meridional Anthills" of the Cape Yorke Peninsula, The proceedings of the royal society of Queensland, vol. 12, pp. 99–100. https://doi.org/10.5962/p.351272. https://www.biodiversitylibrary.org/permissions/. Accessed 2021.

14. American Institute of Physics—Climate control in termite mounds. (2014). Available at: https://www.sciencedaily.com/releases/2014/11/141125074646.htm. Accessed 2021.
15. Turner, J. S., & Soar, R. C. (2008, May 14–16). Beyond biomimicry: What termites can tell us about realizing the living building first International conference on industrialized, intelligent construction (I3CON) Loughborough University.https://www.researchgate.net/publication/255650482_Beyond_biomimicry_What_termites_can_tell_us_about_realizing_the_living_building. Accessed 2022.
16. King, H., Ocko, S., & Mahadevan L. (2015, September 15). Termite mounds harness diurnal temperature oscillations for ventilation. *PNAS 112*37 11589–11593., https://www.pnas.org/doi/full/10.1073/pnas.1423242112
17. Noirot, C., & Darlington, J. (2014). *Termite Nests: Architecture, regulation and defence, chapter of termites: Evolution, sociality, symbioses, ecology* (p. 121). Springer Dordrecht. https://link.springer.com/chapter/10.1007/978-94-017-3223-9_6. Accessed 2023
18. https://asiasociety.org/korea/ondol-korean-traditional-heating-system
19. Bean, R., Olesen, B. W., & Kwang, W. (2010, January). History of radiant heating & cooling systems. *ASHRAE Journal*, 40–47.
20. https://www.japanlivingguide.com/living-in-japan/culture/kotatsu/
21. https://radonna.biz/blog/kotatsu/
22. Schwamborn, I. Y., Comparison of the "Hori-gotatsu" in the traditional Japanese house and the "Kürsü" in the traditional Divri i house, Intercultural Understanding, 2019, 9, pp. 21–25, itcs. mukogawa-u.ac.jp, Accessed 2024.
23. https://content.meteoblue.com/en/research-education/educational-resources/meteoscool/general-climate-zones. Accessed 2023.
24. https://www.britannica.com/science/Koppen-climate-classification. Accessed 2023.
25. Marcus Vitruvius Pollio, de Architectura, Book VI, Chapter 1 https://penelope.uchicago.edu/Thayer/E/Roman/Texts/Vitruvius/6*.html. Accessed Jan 2024.
26. Vitruvius, The Ten Books On Architecture, translated by Morris Hicky Morgan, 1914, Harvard University Press, The project Gutenberg eBook of ten books on architecture, by vitruvius, Release Date: December 31, 2006 [EBook #20239], Produced by Chuck Greif, Melissa Er-Raqabi, Ted Garvin and the online distributed proofreading team at http://www.pgdp.net. Accessed Jan 2024.
27. Silvi, C. (2007). The Italian National Solar Energy History Project. *Proceedings of ISES World Congress, 1*, 3065–3069.
28. Kryza, F. T. (2003). *The power of light: The epic story of man's quest to harness the sun* (pp. 43–46). McGraw-Hill.
29. Dabija, A.-M. (2021). *Alternative envelope components for energy efficient buildings* (p. 64). Springer. https://doi.org/10.1007/978-3-030-70960-0
30. Bagasi, A. A., Calautit, J. K., & Karban, A. S. (2021). Evaluation of the integration of Mashrabiya into the ventilation strategy for buildings in hot climates. *Energies, 14*(3), 530. https://doi.org/10.3390/en14030530
31. Taşkan, D., & Behzad Ismaeel, A. (2022). An architectural element: Mashrabiya/Mimari Bir Eleman: Maşrabiye. *Art-Sanat, 17*, 478. https://doi.org/10.26650/artsanat.2022.17.841296
32. Al-Murahhem F. (2010). The mechanism of the rawāshīn: the case study of Makkah. *Eco-Architecture III, 128*, p. 562. https://www.witpress.com/elibrary/wit-transactions-on-ecology-and-the-environment/128/20814, Accessed Jan 2024.
33. https://www.abiya.ae/knowledge-hiba/mashrabiya-history-and-spread
34. Hossam El-Borombaly, H., & Molina-Prieto, L. F. (2015). Adaptation of vernacular designs for contemporary sustainable architecture in Middle East and Neotropical region, *International Journal of Computer Science and Information Technology Research, 3*, 4, pp. (13–26). www.researchpublish.com, in www.researchgate.net/profile/Hossam-Elborombaly/publication/282660639_Adaptation_of_Vernacular-2292/links/561697e008ae1a8880031381/Adaptation-of-Vernacular-2292.pdf
35. https://passivehouseplus.ie/stack-effect

36. https://www.tecquipment.com/knowledge/2019/daniel-bernoulli-bernoullis-principle-and-equation
37. Saadatian, O., Haw, L. C., Sopian, K., & Sulaiman, M. Y. (2012, April). Review of wind-catcher technologies. *Renewable and Sustainable Energy Reviews, 16*(3), 1477–1495. https://doi.org/10.1016/j.rser.2011.11.037
38. https://www.insideflows.org/project/ancient-wind-catchers-in-hyderabad/
39. A'zami, A. (2005, May). Badgir in traditional Iranian architecture, international conference "Passive and low energy cooling for the built environment", Santorini, Greece, p.1021. https://www.aivc.org/resource/badgir-traditional-iranian-architecture. Accessed Jan 2024.
40. https://knoji.com/article/types-of-traditional-windcatcher-malqaf-part-2/
41. https://earthrivergeothermal.com/history-of-geothermal-systems/
42. https://www.britannica.com/science/geothermal-energy/History
43. https://www.geothermal-dhc.eu/News/Details?id=20
44. Bloomquist, R. G. (2001). *Geothermal District energy system analysis, design, and development*. International Summer School. https://pangea.stanford.edu/ERE/db/IGAstandard/record_detail.php?id=5313

Chapter 5
From Yesterday Toward Tomorrow

Preamble

It is not technology that makes a building special; it is the idea behind the reality of the built structure, materialized in space. In the late 1970s, a Swedish architect and one of the most renowned theoreticians in architecture, Christian Norberg-Schultz, published a book on the phenomenology of architecture [1], nailing the concept of *genius loci*: the spirit of the place. Taking Heidegger's theories to another level—according to his experience and knowledge—he states, from the first pages:

> *"Man dwells when he can orientate himself within and identify himself with an environment, or, in short, when he experiences the environment as meaningful. Dwelling therefore implies something more that "shelter". It implies that the spaces where life occurs are* **places** *in the true sense of the word. A place is a space which has a distinct character. Since ancient times the* **genius loci**, *or "spirit of the place" has been recognized as the concrete reality man has to face and come to terms with in his daily life. Architecture means to visualize the* **genius loci**, *and the task of the architect is to create meaningful places, whereby he helps man to dwell.* [1, page 5]"

Simply beautiful and true. Equipment doesn't really make a building iconic, not even comfortable or welcoming. It just makes it competitive from a quantifiable point of view: energy efficiency, carbon footprint, greenhouse gas emissions, and who knows what other parameters of performance that will be considered in the future. The immateriality of the material is not expressed in the number of BIPV or in the percentage of reclaimed products, it is that *je ne sais quoi* that makes one love a place, remember it, and wish to return to it. Norberg-Schultz defines it: "whereas "space" denotes the three-dimensional organization of the elements which make up a place, "character" denotes the general "atmosphere" which is the most comprehensive property of any space. [...] Similar spacial organisations may possess very different characters, according to the concrete treatment of the space-defining elements (the **boundary**)" [1, page 11]. In respect to the character of a building or a landscape, Norberg-Schultz concludes that it "depends upon, and is therefore

A.-M. Dabija, *Architectural Design Strategies for Saving Energy in Buildings*,
https://doi.org/10.1007/978-3-031-73541-7_5

determined by the technical realisation ("building"). Heidegger points out that the Greek word meant a creative "re-vealing" (**Entbergen**) of truth, and belonged to poiesis, that is "making"" [1, page 15].

> *"Hölderlin was right when he said: "Full of merit, yet poetically, man dwells on this earth". This means: man's merits do not count much if he is unable to dwell **poetically**, that is, to dwell in the true sense of the word. [...] Only poetry in all its forms (also as "the art of living") makes human existence meaningful, and **meaning** is the fundamental human need.*
>   *Architecture belongs to poetry, and its purpose is to help man to dwell. [...] The basic art of architecture s therefore to understand the "vocation" of the place. In this way we protect the earth and become ourselves part of a comprehensive totality. [1, page 23]."*

What an extraordinary beautiful definition of sustainability, in its profound meaning! And how fresh the text is, although 45 years have passed over it.

Buildings represent a distinct environment that encapsulates the level of knowledge of the time when they are built. However, in the effort of building more energy efficient, the traditional/perennial principles—that are applied in building physics—should not be forgotten in favor of the preferential use of technologies.

nZEB avant la lettre

Indoor comfort is not a contemporary requirement. Providing the appropriate air temperature, ventilation, and light was an issue that was addressed, hundreds of years ago too, mainly by innovative design. Even when new technologies were developed, innovative design made the difference in integrating new systems, with consequences in the operating costs of the building(s). However, when technology prevails over the relation with the natural and anthropic environment—it happened in the 1920s with the air conditioning industry and it happens today, one century later with a large range of renewable technologies—the quality of the building space (and architecture itself) suffered.

The quantifiable well-being of the occupants, the materials, the indoor air, and light and sound quality can be expressed in the energy that was/is used to meet the parameters of comfort in the built space. A wise approach is to use the force and the rules of nature, thus reducing some of the operation costs; a less wise approach is to rely only on providing the necessary equipment to bring the space to the required physical conditions.

According to Klepeis et al. [2] in the mid-1990s, people in the United States and Canada were spending over 90% of their time in enclosed spaces. The probability that these figures have changed and that humans are spending more time outdoors than in the 1990s is very small, considering that the population became even more indoor related than 40 years ago: most gym halls and chains opened in the 1980s or later. In this time interval, the building facilities have evolved, and the conditions for achieving indoor comfort can be fulfilled by pressing one button on the remote control. However, quality indoor air was provided prior to the development of the HVAC systems, with a lower—if at all—energy consumption.

nZEB is the acronym of "nearly zero energy buildings," and its deeper meaning is that these buildings use less energy originating from nonrenewable sources. Within certain limits, there are buildings, some more than 100 years old, that, even according to today's comfort requirements, use less energy from conventional resources.

This chapter will present some of the buildings that represent examples of innovative architectural design that uses passive strategies for daylighting and natural ventilation, hence diminishing the needs for traditional energy sources. In some of the case studies, alternative systems were used to fill out the energy demand required for the functioning of the building facilities.

One common denominator of these case studies is that the final goal of the architects was to create people-oriented environments using contemporary tools, which include the use of the "smart" materials of their times and alternative energy sources.

In other words, the comfort and well-being of the occupants was always the main target.

In all the cases—old or new—the careful observation and understanding of nature led to innovative architectural solutions.

Steiff Factory, Giengen, Germany
Coordinator Richard Steiff (1903–1908)

In 1880, Margarete Steiff founded a plush toy factory in Giengen, Germany. "As in her childhood she contacted poliomyelitis,she had her legs paralyzed as well as pain in her right arm". However, ambitious and determined, she managed to go to needlework classes, became a seamstress, and opened a store in 1877 where she sold self-made garments and household articles [3]. Eventually, she managed to hire employees. Sometime around 1880, she found a pattern for a fabric toy elephant designed to be a pincushion and sew it; made of felt and stuffed with mohair, the "Elefäntle" was highly appreciated by children, and therefore, instead of a pincushion, it was successfully sold as a toy. In 1893, Margarete Steiff founded a felt goods factory [3].

In 1897, Richard Steiff, Margarete's nephew, joins the company. He was a trained artist who graduated from the School of Applied Arts in Stuttgart and also studied in London. His talent and expertise led not only to the diversification of the toy production but also to surprising inventions in the technology of buildings.

Richard Steiff is the creator of the teddy bear,[1] the first jointed[2] toy animal. In the building industry, he is the coordinator of the design of one of the first curtain wall

---

[1] "In 1902 Richard Steiff, Margarete's nephew, designed a toy bear made of mohair called Bear 55 PB.

Inspired by American President Theodore Roosevelt, the initially nameless bear received its name Teddy, which is still known today. Roosevelt refused to shoot a tethered bear during a hunting trip. The incident was captured by cartoonist Clifford K. Berryman and published in the Washington Post.

This was the best publicity for our Teddy bear—the teddy boom began and the Steiff brand achieved worldwide recognition" [4].

[2] Articulated dolls were made in the Greek and Roman Antiquity, using wire connections [5].

buildings (the first one in Germany): the Margarete Steiff toy factory built in Giengen in 1903.

In order to ensure better working conditions for the labors (usually women) without engaging higher maintenance costs (heating, lighting), Richard Steiff's design of the facades included two fully glazed walls, so that the workers would benefit to the maximum from the use of daylight in the workspaces and, at the same time, have a better protection against the strong winter winds. The result was a three-storey building (storage on the ground floor, workspace on the floors) with continuous glass skin. Furthermore, it is the earliest example of a curtain wall in Germany [6] and the first double-skin curtain wall in history.

The workers in the toy factory were mostly women; hence, the building was known as the "Jungfrauenaquarium" (Young Women's Aquarium). Today, it is known as the East Building.

The exterior glazing of the Steiff factory is continuous, while the interior glass runs from one storey to another. The two glass skins form a buffer zone that provide a better thermal insulation in winter. By doing so, heating costs were significantly reduced. To avoid overheating during the summer season, the glass was painted with lime paint in spring (that reflected the solar radiation), paint that was washed away by rain in the autumn. Operable windows allow for natural ventilation of the workspaces. The large glazed surfaces allowed the natural light to flow in, reducing the need for artificial light and the costs, accordingly (Fig. 5.1).

Fig. 5.1 Building of the Steiff "Jungfrauenaquarium," 1903. Unknown München architects. (Photo: Holger Baschleben, CC BY 2.0. Source: https://www.flickr.com/photos/holger-baschleben/5007435409/in/photolist-D8JLK-8Cuq6g)

As the steel and glass assembly was not airtight (the slim steel components and the glass panes allowed a slight air movement), some faible ventilation could take place, and condensation was improbable between the two glass skins. The building functioned on natural ventilation (operable windows) as air conditioning was invented in 1925[3] and introduced in buildings, on a large scale, after the Second world War.

To enable Margarete to access each of the production floors by wheelchair, Richard built a generous exterior ramp [6], inventing the ramp[4] for disabled persons.

In the next years, as the demand of the teddy bears increased, two more buildings were added: in 1904 and in 1908. The concept of the constructive system was preserved, but the structure was changed from steel to wooden columns. All three buildings can still be seen and are listed as historical monuments.

Ideas of double-skin facades seem to have been launched much earlier. According to Dirk Saelens, cited in several publications [8] as early as 1849, "Jean-Baptiste Jobard, at that time director of the industrial Museum in Brussels, described an early version of a mechanically ventilated multiple skin façade. He mentions how in winter hot air should be circulated between two glazings, while in summer it should be cold air."

Casa Batllo, Barcelona, Spain
Arch. Antoni Gaudi, 1904

It was, in fact, a renovation of an existing building owned by the Batllo family. Built in 1877, when there was no electric light in Barcelona, Josep Batlló y Casanovas, a prominent businessman, commissioned Gaudi to modernize the house (that originally was due to be demolished) [9]. The result was not only an integration of the building facilities within the existing edifice but an example of innovative design approach that also provided natural light and ventilation, without additional costs (Fig. 5.2, left).

The principles used by Gaudi are the same that had been used in traditional housing and will be used by contemporary architects; the results are, technically speaking (and according to the scale of the buildings), similar; yet, the volumetry and the images are different as the environment, materials, techniques, equipment, and— above all—the architects' imagination differ.

Today the building is a UNESCO Heritage.

Casa Batllo can be studied from several angles, its innovative design remaining fresh and powerful one century later. The "facettes" that will be emphasized in the following paragraphs will only be related to improving the quality of the indoor environment, through light and ventilation, as the consequences can be measured in energy savings provided by the ingenious design.

---

[3] Willis Carrier is the engineer who launched the principle of commercial air conditioning [7].

[4] It seems that ramps existed in the ancient Greece, but, as many other inventions, they were forgotten and "rediscovered" in the Giengen building, at the beginning of the twentieth century.

The edifice is located on the Paseo de Gracia, a street that became, in the nineteenth century, the main promenade of the city and later the avenue of the cars. It offers a short facade (14.5 m^5 against almost four times as much, in depth) toward the street and is flanked by three other representative buildings of the 1900 Catalan architectural style.[6] In fact, due to the antagonic architecture, the four buildings form a block that is called "Illa de la Discòrdia" (Fig. 5.2, right).

Fig. 5.2 Architect Antoni Gaudi i Cornet Casa Batllo, Barcelona. (Photo Ana-Maria Dabija)

Gaudi convinced the owner (who intended to keep the "*piano nobile*" for his family and rent the other floors) to build an inner courtyard—a large shaft—even if that meant that less habitable space could be rented. The shaft/courtyard is the central piece of a coordinated natural light and ventilation system that connects the facades with the rooms and eventually evacuates the air into the shaft(s). It is in fact a cross-ventilation system. The principle was introduced at the scale of large buildings, more than 100 hundred years later, as the atrium of office buildings.

---

[5] https://www.barcelona-tickets.com/casa-batllo-tickets/casa-batllo-architecture/

[6] The architects of the four "discording" buildings are Lluís Domènech i Montaner, Antoni Gaudí, Josep Puig i Cadafalch, and Enric Sagnier.

The façade has an interconnected, accessible, and adjustable system of apertures in the wooden segments (Fig. 5.3, right) as well as sliding, operable windows, allowing the tenants to open or close them (manually) according to the season (Fig. 5.3, left).

Fig. 5.3 Casa Batllo. Air intake from the façade. (Photos Ana-Maria Dabija)

The doors on the trajectory to the inner courtyard are also provided with apertures, leading the air toward the great shaft from where it is exhausted outside (Fig. 5.4, right).

Fig. 5.4 Casa Batllo. Parapet apertures toward the interior courtyard/shaft. (Photo Ana-Maria Dabija)

The central courtyard (Fig. 5.4, left, Fig. 5.7) is not the only shaft in the building: ventilation grids and operable louvres are the discrete elements that certify the existence of small dimension shafts, hidden behind walls (Fig. 5.5)…

Fig. 5.5 Casa Batllo. Ventilation apertures. (Photo Ana-Maria Dabija)

…and, of course, the sculptural, decorated chimneys (Fig. 5.6) that rise above the terrace prove the existence of hidden ducts.

Fig. 5.6 Casa Batllo. Chimneys. (Photo Ana-Maria Dabija)

Apart from being an impressive (at the scale of the building) component of the natural ventilation system, the interior courtyard has the role of providing daylight in the areas where, due to the depth of the building, the sun would not have entered. A master of the art of detail, Gaudi used the transparence, reflexivity, brightness, and color of the materials to maximize the daylight and minimize the glare. At the upper levels of the house, the window dimensions decrease (Fig. 5.7, left) although their geometry is preserved (giving also a sensation of height but this is not relevant to the current topic); the ceramic tiles are intense blue at the top of the shaft (Fig. 5.7, right), and the color fades to a lightshade on the lower floors (Fig. 5.7, left, Fig. 5.4, left) to provide the appropriate daylight in the interior spaces by absorbing excessive sunlight on the top floor and by increasing the reflections on the lower floors, where direct natural light was less.

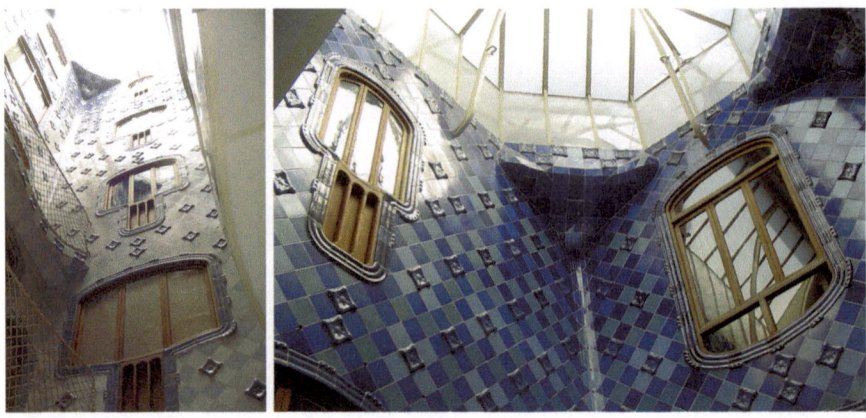

Fig. 5.7 Casa Batllo. Daylight solutions by window geometry and finishing materials. (Photo Ana-Maria Dabija)

All Gaudi's buildings, culminating with the unfinished church of the Sagrada Familia, testify his preoccupation of observing, studying, understanding, learning, and applying the ever-lasting rules of nature with breathtaking and surprising results. He is the precursor of the biomimicry and parametric architecture which, both, have their roots in the surrounding natural world. However, the natural world only teaches those who see and accept the lessons.

From Yesterday Toward Tomorrow

For more than three decades, the international legislation has been oriented toward carbon reduction and energy efficiency in all lines of business.[7]

Buildings did not make an exception as they are a finished product themselves and at the same time host all the human activities that are carried out in an appropriate—mainly man-made—environment. Therefore, buildings are the expression of what we are **now**. However, "now" is not synonym with "new." Historians, philosophers, and writers state that the past and future are connected; Victor Hugo wrote that the future is a door, and the past is the key.[8]

The same philosophy applies to building design.

---

[7] See Chap. 2 for more information.

[8] *L'avenir est une porte, le passé en est la clé* in https://citations.ouest-france.fr/citation-victor-hugo/avenir-porte-passe-cle-121125.html

Queen's Building, De Montfort University, Leicester, UK
Architecture: Short Ford Associates—Alan Short and Brian Ford
Services Consultant: Max Fordham Associates
Year of Completion: 1993

In 1989, De Montfort University in Leicester "set out to design Europe's largest naturally ventilated building" [10, 11]. Alan Short and Brian Ford were the architects who designed, interpreting the timeless principles of building physics and with the collaboration of an interdisciplinary team of specialists (including academic scientists), managed, and built an emblematic building (Fig. 5.8) that can be considered an example of architecture where shape volume and materials are a consequence of the response to the natural forces—air movement, light, and heat.

Fig. 5.8 Architecture: Short Ford Associates—Alan Short and Brian Ford—Queens Building, De Montfort University 1993. Steve Cadman CC BY-SA 2.0. (Source: https://www.flickr.com/photos/stevecadman/47963797)

Based on "understanding how modern perceptions of safe and comfortable environments evolved through early understanding of disease propagation through the air," he observed that nineteenth-century hospital designs proved to be innovative and represented "proto-modern buildings," suggesting that

> *"aggressive mid-Twentieth Century advertising of air conditioning killed a highly productive stream of architectural design, overdue for vigorous reexamination to shift the prevailing 'will to form'. [...] given the fear of bad air, a sustained campaign from the later eighteenth century was uncovered to make safe, naturally ventilated environments in public buildings. The stakes couldn't have been higher, the potential rewards imagined to be the guarantee of public health, longed for social recognition and a possibility of tremendous financial return. The campaign became increasingly bullish in its scientific content, pooling evidence and understanding of fundamental physics, fluid dynamics, chemistry and human biology. [12, page 330]."*

As the industry of heating and ventilating began to flourish, the possibility of rapidly providing a comfortable indoor environment became accessible for an increasing number of people in the United States and then in Europe: "This was perceived to be a huge social good but, we now know, at an immense cost in energy and carbon and ironically, perhaps, in wellbeing" [12, page 330]. "Without an appreciation of both"—the poetic imagination and the understanding of how the natural forces act in what is defined scientifically as buildings' physics—"one is left in the hands of the air-conditioning industry whose simple aim, not unreasonably, is to capture market share and generate profit, or a gestural architecture with no environmental presence or meaning [12, page 332].

The Queens Building of the De Montfort University represents one of the examples of bridges thrown between past and present complex design approaches, and, due to the periodical assessment of the building, lessons have been learned, and improvements of the design strategies (as some passive systems are enhanced through active means) have been crystalized.

The goal was not to promote old technologies but to design buildings that, by using low-energy techniques, would substantially reduce the carbon dioxide emissions[9] (the unit of measurement that compares, horizontally, from a transdisciplinary perspective, the energy consumption) while maintaining the same level of thermal, visual, light, and acoustic comfort and the appropriate indoor air quality.

---

[9] According to the official European definitions stipulated in [13], "carbon dioxide emissions or CO_2 emissions are emissions stemming from the burning of fossil fuels and the manufacture of cement; they include carbon dioxide produced during consumption of solid, liquid, and gas fuels as well as gas flaring."

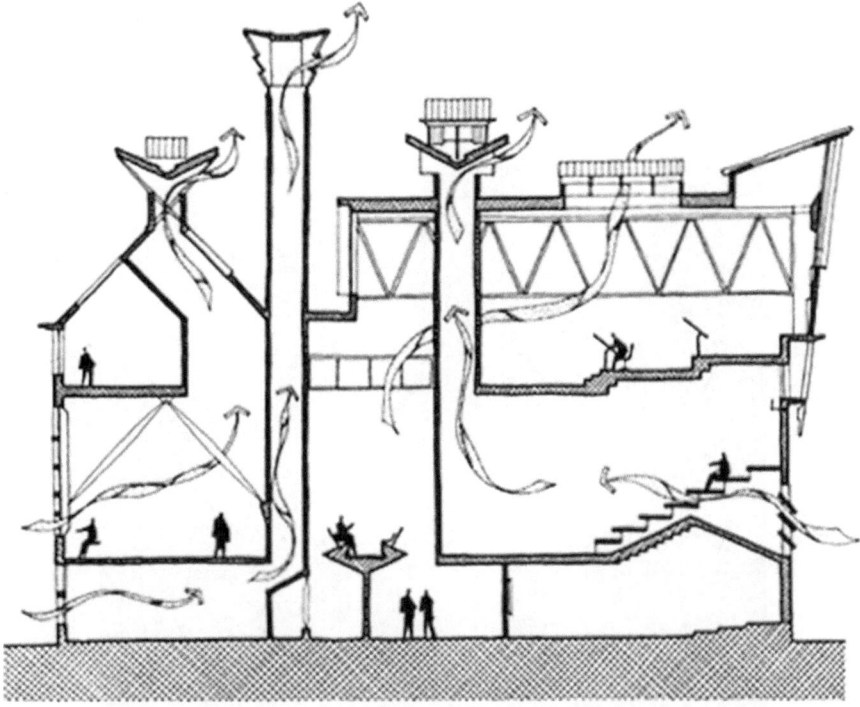

Fig. 5.9 Queens Building, De Montfort University, cross-section illustrating the natural ventilation of the spaces of the building, from the inlets and dampers to the chimneys. (Image: by courtesy of Professor C. Alan Short)

In a nutshell, the main passive leverages that the architects used in Queens Building were as follows (Fig. 5.9):

– Thermal insulation using massive brick walls (and their thermal mass)
– Daylight provided by generous windows and skylight
– Cross-ventilation—whenever applicable—and controlled air movement through inlets, dampers, stacks, and chimneys

The result was a building that provided good, healthy indoor conditions using less energy.

Commerzbank Headquarters, Frankfurt
Architecture: Foster + Partners
Year of Completion: 1997

"At fifty-three storeys, the Commerzbank is the world's first ecological office tower and on completion it was the tallest building in Europe. The project explores the nature of the office environment, developing new ideas for its ecology and working patterns. Central to this concept is a reliance on natural systems of lighting and ventilation. Every office is daylit and has openable windows, allowing the occupants to control their own environment. The result is energy consumption levels equivalent to half those of conventional office towers— the offices are now naturally ventilated for 85% of the year. [14]"

Fig. 5.10 Architecture: Foster and Associates, Commerzbank Headquarters. (Photo: Chaouki, CC BY-SA 2.0. Source: https://www.flickr.com/photos/weltum/468475529)

The first paragraph of the presentation of the building, on the Foster and Partners site, defines its features: a construction where daylight, air movement, and the use of vegetation are the main instruments for achieving indoor comfort (Fig. 5.10). In other words, for creating architectural space by using nature in the benefit of the building occupants (office workers in this case) with a minimum of energy use in operation. One of the quantifiable results is "energy consumption levels equivalent to half those of conventional office towers—the offices are now naturally ventilated for 85% of the year" [14].

The building plan is triangular (Fig. 5.11) with the service corps that occupy the building edges and the sides assigned to office spaces and winter gardens. The central part is a full-height central atrium that acts as a ventilation shaft.

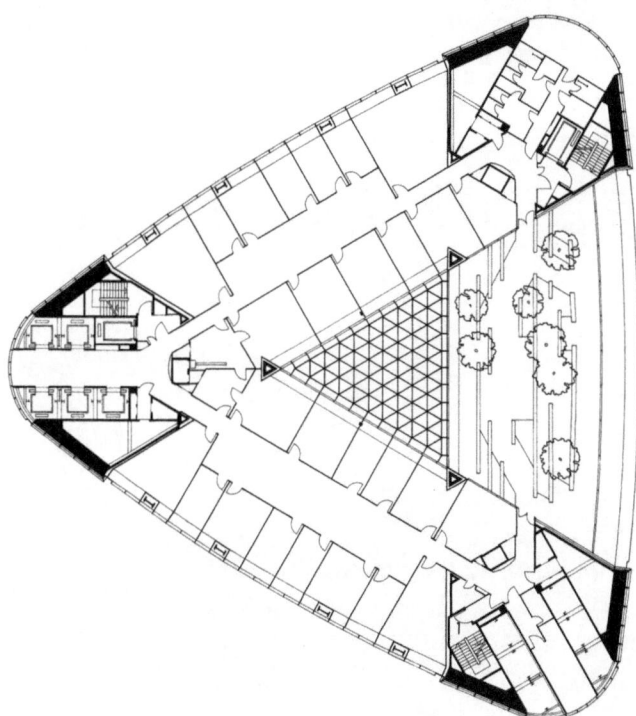

Fig. 5.11 Commerzbank floor plan. By courtesy of Foster + Partners

"Storey-clusters" consisting of packs of four office floors and three-storey high winter garden rotate by 120° against the previous, with the cycle repeating and giving the image of a spiral climbing toward the top of the building. "From the outside these gardens in the sky give the building a sense of transparency and lightness. Socially, they form focal points for village-like clusters of offices, providing places to meet colleagues or relax during breaks. Environmentally, they bring light and fresh air into the central atrium, which acts as a natural ventilation chimney for the inward-facing offices" [14]. Skylights divide the atrium (Fig. 5.12) into three sections, providing daylight and air movement control in the atrium (detailed analysis showed that if the atrium had not been sectioned, the strong air currents would have been hard to control) [15].

Fig. 5.12 Downview through the atrium. (Photo: © Nigel Young/Foster + Partners. By courtesy of Foster + Partners)

The facades of the winter gardens (Figs. 5.12 and 5.13) are equipped with operable, double-glazed panels, allowing fresh air to enter the building; toward the atrium, they are fully open and represent areas for networking, socializing, leisure. All the interior windows facing the gardens can be opened, thus supplying fresh air in the offices.

Fig. 5.13 View of a winter garden. (Photo: © Jens Willebrand. By courtesy of Foster + Partners)

More than having just a social role and an important contribution to the health and well-being of the occupants, the spaces of the winter gardens cover several air comfort issues:

– The incoming air is pre-heated due to the passive gain of the glazed surface; hence, the winter gardens act as buffer zones.
– The incoming air is filtered by the plants; in consequence, the air that enters the atrium and offices has a higher purity.
– Through the fully glazed "greenhouses," the natural light penetrates at any angle throughout the whole year, while the reflections of the glass prisms (the offices and the atrium) transmit it further in the deepest corners (Figs. 5.12 and 5.13).
– The sky gardens are a component of the cross-ventilation system (with the climate facade and the central atrium shaft) for all the office areas (Fig. 5.14).

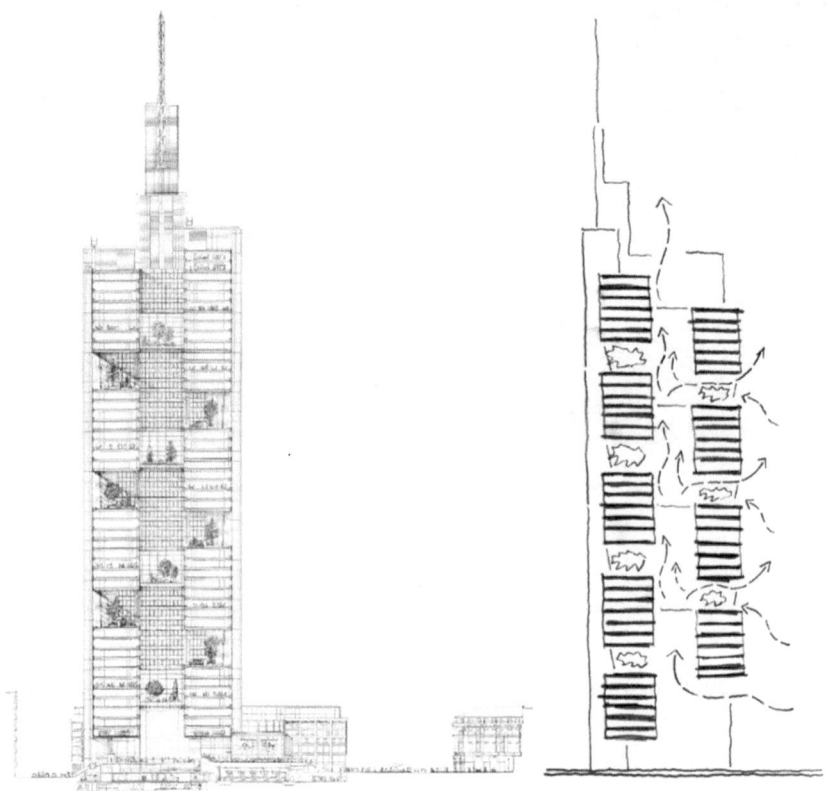

Fig. 5.14 Cross-section and sketch illustrating the natural ventilation principle. (Image: By courtesy of Foster + Partners)

The "Klimafassade" (climate façade) is a double-skin curtain wall, with an exterior fixed laminated glass pane and interior tilting windows provided with insulated glass units. The exterior and the interior glazing form a ventilated air cavity connected, through intake and exhaust slots, to the outside—hence, air flows between the two "skins" and is supplied in the interior, when the windows are opened; mechanical operated blinds are installed behind the exterior glass pane, in the air cavity.

When the windows are closed, mechanical ventilation is provided.

The heating of the space is ensured with perimetral devices, while cooling is accomplished with chilled ceilings.

A sophisticated BMS was installed, with sensors that monitor the temperature, humidity, and light level and switch on or off the systems according to the occupancy of the offices, thus decreasing the unnecessary use of energy. The manual access to operating the windows is blocked when the outside temperature falls beyond the established values that allow natural ventilation.

Over the first decade in operation, the data regarding the measured energy consumption showed that this building consumes less than half compared to equivalent towers where passive strategies were not included and that the building is naturally ventilated throughout 85% of a year. From January 1, 2008, the Commerzbank HQ is exclusively supplied from renewable energy, turning it into a net zero carbon building [16].

Cultural Center Jean-Marie Tjibaou, New Caledonia
Architecture: Renzo Piano Building Workshop, Architects; Ventilation and MEP Engineering Concept: Ove Arup & Partners
Year of Completion: 1998

New Caledonia is an archipelago situated in the Southern Hemisphere, in the southwest Pacific Ocean. The climate is tropical, with temperatures that vary between 18°C in the "comfortable and mostly clear[10]" winters and 30°C in the "warm, oppressive, wet, and partly cloudy[11]" summers. The annual mean relative humidity is about 75% . The dominant wind blows from the southwest, but cyclonic winds can gust up to approximately 65 m/s from any direction [17].

New Caledonia's ecosystems include several vegetation types, including evergreen rainforests, making it one of the richest diversity per square meter on the planet [18].

Following an international competition and an invitation-only tender in 1991, the construction of this exotic assembly began in 1993, in Noumea, New Caledonia. "The project had to be designed to honour traditional Kanak culture, while at the same time providing a focal point for the inevitable development of its society. [...] the objective was to ensure that, despite its adaptations, the Kanak culture would not lose touch with its historical roots" [19].

Aiming to connect to the local tradition[12] in a contemporary interpretation, the assembly "grows" organic in the landscape, between the tall trees, resembling them as shape, texture (materials), and transparency: one can see through the "shells" as through the branches of the surrounding trees (see Fig. 5.15).

---

[10] https://weatherspark.com/y/144754/Average-Weather-in-Noum%C3%A9a-New-Caledonia-Year-Round

[11] ibidem.

[12] For this project, the architect turned to an ethnologist, throughout the design process, from the competition to the final phase.

Fig. 5.15 Architect Renzo Piano, The Cultural Center Jean-Marie Tjibaou, Nouméa, 1998. (Photo: Bruno Moure, CC BY 2.0. Source: https://www.flickr.com/photos/brunomoure/20816304962)

The location of the Center is "between the open sea and the protected lagoon, is set against the backdrop of the mountains and the promontories jutting out into Magenta bay" [19], in a landscape where the luxuriant vegetation, the water, the mountains, and the sky intertwine.

Consistent with the intent of integrating into the local atmosphere, the Center expresses discretion and respect toward the culture and natural and man-made environment while exposing a bold and contemporary view: 10 "huts," of different dimensions and heights, are scattered in the natural scenery like the houses of a local village (Fig. 5.15). The architectural expression of the wooden "shells" suggests (but not replicates) the Kanak hut construction and creates a dialogue between the local culture and materials—wood and stone—and the contemporary world, with the specific, multiple functions expressed with contemporary means, glass, aluminum, steel, and lightweight technologies, along the paths and between the natural and created voids of the Center.

The "shells" ("huts," "houses," "scoops," "cases" as they were named in the scientific literature) consist of laminated timber structures that cover about 2/3 of a circle. The outer wall structure consists of arches that give the curved shape of the units, while the inner wall is made with vertical columns. The two concentric walls form a stiff structure, connected with metal diagonals and bracings to provide structural resistance and to resist the strong winds and cyclones. Iroko wood was selected for these elements, although it is not a local species, it is "a stable wood that's resistant to termites and can even be used as a laminate spaces created are endowed with rather different characters based on their intended purposes" [19]. The cladding of the double "shell" is made with hardwood of different widths, spaced unevenly, allowing the air circulation between the exterior and the interior. At the upper part, the cladding of the two twin structures defines a "chimney." The natural air ventilation system relies on the operable windows, the openings, and the chimney. The air passes through the "shell" louvres, spreads through the building, and is exhausted through the curved chimney, according to the wind direction and intensity: as the wind speed increases, the louvres close, from bottom toward the top (Fig. 5.16).

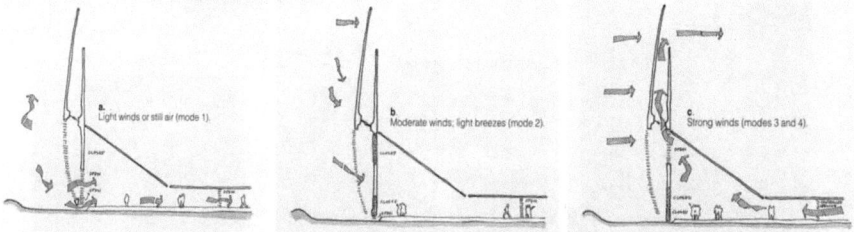

Fig. 5.16 Natural air ventilation strategy for still air (**a**), light breezes (**b**), and strong winds (**c**), according to [17]. (Images: © ARUP, by courtesy)

"Several openings in the Case allow ventilation. Two sets are located at the front (facing prevailing winds): 2 m high openable louvre windows and, at low level, 0.5 m high openable louvres. These openings are controlled automatically to either fully-open or fully-closed positions. At the back of the Case a series of openable windows allow cross-ventilation. [...] The chimney has several functions. During days with little or no wind, it helps induce ventilation by stack effect or natural convection. During periods of high wind speeds, the shape of the shell directs air up and across the chimney inducing negative pressure. This then draws air through the internal space from openings at the back of the Case. [...] [17]."

The double roof (Fig. 5.17) allows air circulation between the two covering layers of wood laminate. An insulated panel with external shading is provided, to reduce surface temperatures. Where needed, wooden blinds protected the transparent glass against the sun.

Fig. 5.17 Architect Renzo Piano, The Tjibaou Cultural Center 1998. (Photo: Gérard, CC BY-SA 2.0. Source: https://www.flickr.com/photos/35803445@N07)

The system was verified and adjusted by computer simulations from the early design stage.

BedZED, Sutton, South London
Design Director: Bill Dunster and ZEDfactory
MEP Engineers: Arup
Sustainability Consultants: Bioregional
Year of Completion: 2002

The adoption of the Kyoto Protocol in 1997 can be considered the background and the trigger for this unprecedented experiment: the Zero (fossil) Energy Development in Beddington, London. In other words, BedZED. By coincidence (or not), the year of completion of the assembly is also the year when the EU Directive on Energy Performance in Buildings was signed. The whole team of specialists—experienced in designing with respect for the environment (Bill Dunster built his house "testing" some of the theories implemented in BedZED [20])—was determined to prove that a holistic design approach can lead to buildings where life can be led at the contemporary standards while saving resources and, in consequence, at low(er) energy costs: "Bill Dunster, collaborating with engineers Arup and Bioregional, was looking for an opportunity to create a zero fossil fuel eco-village. Peabody Trust, a long-established, London-based housing association, was recruited as the enlightened developer for this site. It was interested in how sustainable living could cut energy and water bills for its low-income tenants" [21].

The aim was to provide good quality space for all the accommodated functions: affordable or social housing, offices, and childcare and community spaces (Fig. 5.18); in general, "it envisioned a more comprehensive strategic thinking, encompassing the environmental, social and economic dimension of sustainability" [22], with appropriate solutions that took into account nature in all its features:

(a) As a "partner in design," expressed in respecting the principles of building physics
(b) As resource for building materials with low embodied energy but with high potential of thermal mass and insulation
(c) As promoter and integrator of new technologies that use no fossil fuels

(a) "Building with nature[13]" means designing in respect to the environment and using the rules of nature, as expressed by the building physics for providing comfort conditions that include thermal, air, light, and visual comfort. Hence, the volumetry and space organization of the buildings are decided by understanding the site: cardinal orientation, sun path, and air movement.

Fig. 5.18 Architecture: Bill Dunster and ZED Factory BedZED aerial view. (Photo: Bioregional CC BY 2.0. Source: https://www.flickr.com/photos/bioregional/10204276634)

---

[13] As defined in other chapters of this book as well as in [23].

Workspaces are located on the northern side of the assemblies, benefiting of even daylighting and reduced glare. The domestic spaces are south orientated, receiving sunlight through generous windows and winter gardens (Fig. 5.19). Thus, the need of artificial lighting is reduced and the energy costs are, in consequence, lower.

The building envelope consists of 30-cm-thick thermal insulation, placed in the three-layered exterior walls (composed of bricks, mineral wool filling, and concrete blocks), and in the ground-level concrete floors and roofing which together with the performant triple glazed windows (argon filled and with low-e coating) and the buffer spaces of the winter gardens were the only designed sources for warming the rooms (Fig. 5.19). The designers questioned "the need for conventional room space heating, in which a system is simply sized to provide comfort, with regulation minimum thermal insulation. Yet many buildings have internal heat gains from people and their activities. So why not size the insulation, with thermal mass heat storage, so that this heat is sufficient to provide space heating through day and night, thus avoiding the need for any conventional heating?" [20, page 12].

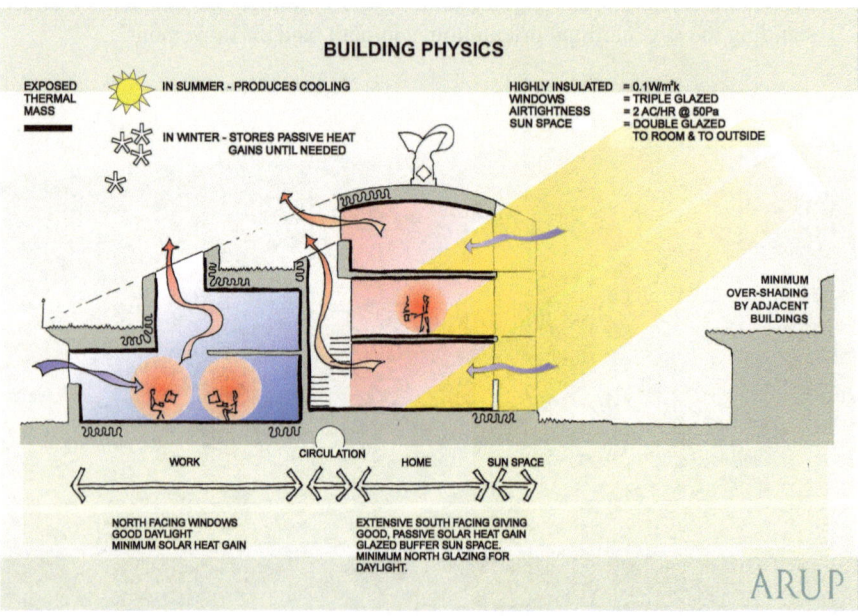

Fig. 5.19 N-S cross-section; diagram of the passive design strategy for natural ventilation and solar approach. (Image: ©Arup, by courtesy)

While air tightness is sought in winter to prevent thermal loss, cross-ventilation, stack effect, and thermal mass combined with the functioning of the multicolored cowls ensure the natural ventilation of the workspaces as well as the habitable spaces.

The costs for heating/cooling of the habitable space are significantly lower as less energy is required.

(b) Most of the construction materials have low-embodied energy and are available within a 35-mile radius of the site. Heavier building materials, like structural steel for the structure of the northern workplaces and reclaimed timber for the partition structure, were brought from a maximum distance of 55 km. Thus, the costs of transport were reduced (during the construction phase), and less embodied energy was used.

(c) A combined heat and power (CHP) system topped up by building-integrated photovoltaic panels was adopted. Unfortunately, it malfunctioned and was replaced by a biomass boiler. The PV panels (Fig. 5.20) were placed and oriented to act as shading devices during the summer while allowing the sun to enter the winter garden in winter, for passive gain [24].

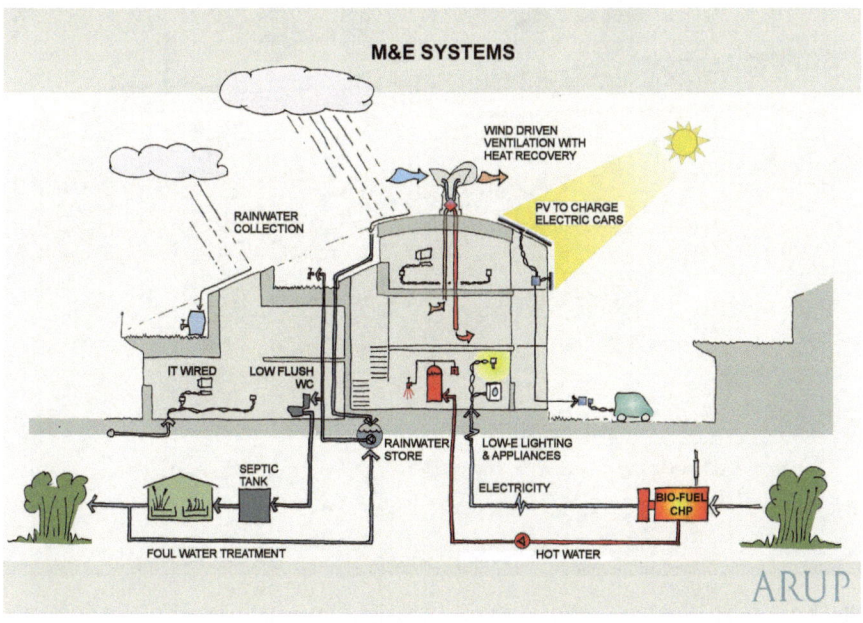

Fig. 5.20 N-S cross-section; diagram of the active design strategy for rainwater storage and treatment, PV and CHP. (Image: ©Arup, by courtesy)

The private gardens are also a source for the general state of well-being of the users—71 of the 89 units have gardens [22]. Harvested rainwater is used as greywater, for watering the plants as well as for other non-potable purposes, hence reducing the water demand (Fig. 5.20).

The feedback was positive, according to *The Arup Journal* [20, page 17]: "This interest has continued to increase. […] The most frequent reason given for wanting to live at BedZED is the modern green lifestyle (63% of occupant survey respondees), with innovative design coming a close second (61%). Popular design features include the sunspace, the gardens, and the sense of space in the homes."

School of Slavonic and East European Studies
Architects: Short and Associates
Year of Construction: 2004

The passive design instruments in Alan Shorts' architecture are the use of thermal mass of the building materials, natural light, and ventilation. However, the architectural expression the quality and the nature of the space, although recognizable as bearing the architect's personal mark, are unique in each building (Fig. 5.21).

Fig. 5.21 Architecture: Short and Associates. School of Slavonic and East European Studies. (Photo N19 ± CC BY-SA 3.0. Source: https://en.m.wikipedia.org/wiki/File:UCL_School_of_ Slavonic_and_East_European_Studies_-_panoramio.jpg)

Based on the experience and the evaluations of the buildings' energy efficiency, carried out through previous projects (like the Queens Building), professor architect Alan Short continued to study, improve, refine, and promote the passive and hybrid architectural strategies and to verify them through the design and construction of other emblematic buildings where the same principles were used, emphasizing the necessity for "a radical return to natural environments in public architecture" [12, Page 328]. In the design of the School of Slavonic and East European Studies (SSEES), he introduced and experimented a new concept in the field of natural ventilation: the passive downdraught cooling.

SSEES has a centenary history: it was founded in 1915; it became an independent institute of the University of London in 1932 and part of the University College London (UCL) in 1999. Located on a site with "personality"—surrounded on three sides by other UCL buildings—the shape of the School was defined as a response to the need of providing the appropriate natural light conditions in both the existing building (Chemistry Building) and the new building [25].

The street-facing thick load-bearing brick facade integrates into the urban landscape but is at the same time one of the instruments that architect C. Alan Short adopts in managing the natural ventilation system of the building: it is a part of a double facade with multiple roles: functionally, it accommodates the main stairs; aesthetically, it reinterprets in modern expression the architectural Georgian style of the center of London while being at the same time an important component of the passive design strategy of increasing the thermal insulation of the building, providing natural ventilation (given the thermal mass that has a role in the air buoyancy, hence contributing to the exhaustion of the stale air at the top of the building), natural lighting, and the protection of the interior space against the traffic noise.

The five stories are penetrated by a fully glazed central light well—an authentic *"piece de resistence"* of the building and the element around which everything is organized—with a double role: to bring light in an otherwise dark area of the built space as well as to supply "clean" air in the building (Fig. 5.22).

Fig. 5.22 SSEES library light well. (Photo left: © Tony Slade, by courtesy of UCL Educational Media, 2005. Photo right: © Gillian Long, by courtesy)

The central light well is a "container" and "distributor" of fresh air; it is provided on every floor with operable windows, and from there, through a sophisticated system of air inlets and dampers, the air is circulated in all the spaces and is eventually evacuated through appropriate chimneys and double facade.

The light well has also a dynamic architectural expression: the glass that can be occasionally replaced with colored panes as in an art project, in 2010 (Fig. 5.22, right).

The passive design strategy includes a distinct scenario for the winter and for the summer season:

– In winter, the air is drawn from the front and the sides of the building and is conducted into a 1-m-deep zone at the bottom of the light well that represents the plenum for incoming air; here, it passes through heating coils and enters into the enclosed light well (Fig. 5.23) that acts like a container of fresh, pre-heated, ascendant air.

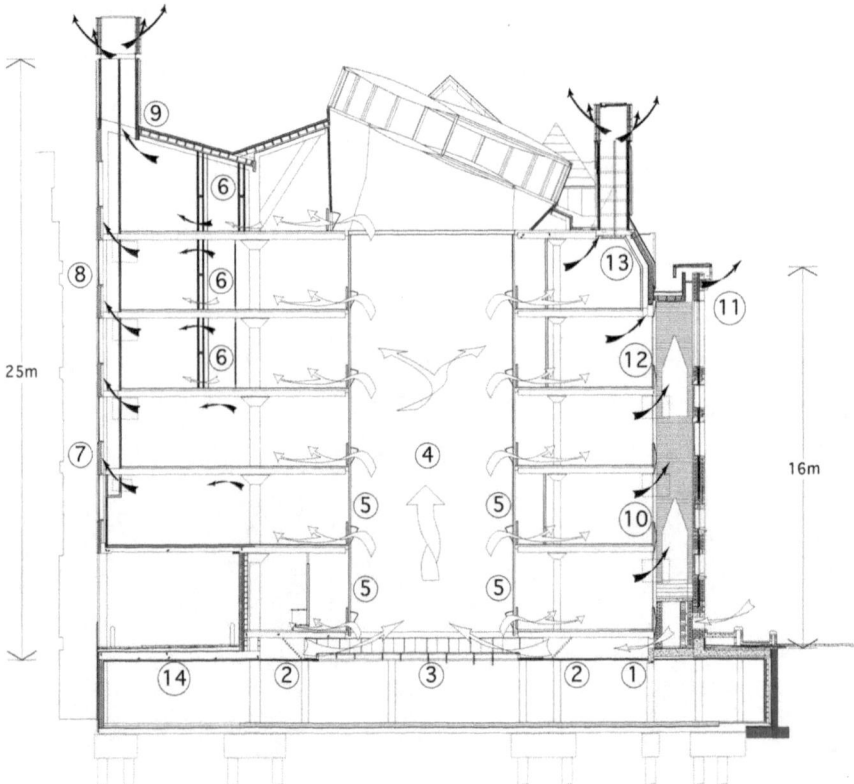

Fig. 5.23 Passive design strategy for winter: the air is heated by the coils ② at the base of the well, raises through the light well ④, and enters in the rooms ⑤, from where it is evacuated through chimneys ⑦–⑩ [25]. (Image: by courtesy of Professor C. Alan Short)

– During the summer, the air is introduced at the upper part of the well (Fig. 5.24), is cooled by coils, and descends into the "air container" from where it is distributed on the floors. The passive downdraught cooling extended the boundaries of the design strategies that use the principles of building physics—thermal mass, passive solar design, and natural ventilation—with the category of descendent air movement. "This low-energy technique enables cooled air to be distributed throughout the building without mechanical assistance. The underlying principles of the technique were explored using physical models and the anticipated performance predicted using thermal modelling" [26].

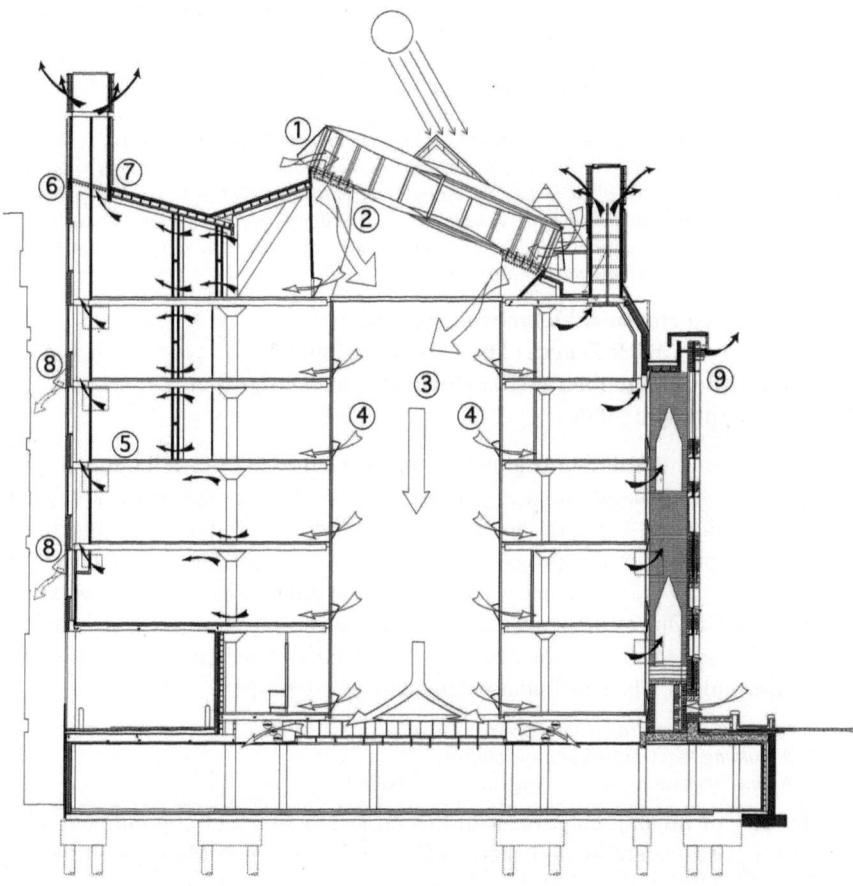

Fig. 5.24 Passive design strategy for summer: the air is cooled by the upper coils ②, descends through the light well ④, and enters in the rooms ⑤, from where it is evacuated through chimneys ⑦–⑩ [25]. (Image: by courtesy of Professor C. Alan Short)

Hence, the system becomes hybrid, as energy from the building facilities network fill in what cannot be accomplished by passive measures [27].

Daylighting and natural ventilation are adjusted by the building's energy management system that monitors the air movement with sensors and computer software (unlike the Casa Batllo where the inlets, grids, and windows are manually operated).

All the hypothesis and theories were studied and verified through computer modelling, simulations, and on-site monitoring, in cooperation with team(s) of researchers specialized in heat and mass transfer processes between gas and liquids and by building facilities engineers, until the whole system was finely tuned:

> *"Stack ventilation systems work effectively when the air in the occupied spaces, especially that below the ceiling soffit, is warmer, and so more buoyant, than ambient air. The physical modelling demonstrated that when the occupied spaces are cooler than ambient, most notably on hot summer days when the downdraught cooling system was in use, the ventilation system could stall. This could lead to elevated temperatures and poor ventilation.*
>
> *To guard against stalling, two design features were included: opening panels at the base level of each stack and the introduction of heat (a by-product of the mechanical cooling system) below the head of each stack (to increase the buoyancy of the air). As the height of the venting parapet, at the top of the double facade, is below the top of the cool air reservoir (in the light well), fresh air should flow freely through the spaces facing Taviton Street, irrespective of the ambient temperature. [25, page 199]."*

"CH2" Council House 2, Melbourne, Australia
Design Director: Mick Pearce (The City of Melbourne)
Design Architects: Stephen Webb, Chris Thorne (DesignInc)
Year of Completion: 2006

The story of this building begins in 2000, when the City of Melbourne decided to build a new headquarter, focusing on providing an interior environment that takes into account the well-being of the occupants: healthy, comfortable, and functional: "The great thing about this project and the way the City of Melbourne have done it—it is a building for the people. …it has been the prime driver through everything no matter what. One of the key design priorities of the CH2 project is to provide a building that is sensitive to the health and wellbeing of the employees" (Stephen Webb, DesignInc). The new building aimed to ensure [28]:

- *"A lighthouse environmental project*
- *A building that was greenhouse neutral*
- *A space that would improve employee well-being"*

The City of Melbourne gave, with this project, a signal of commitment to the principles of sustainability and energy efficiency, confirming Mies van der Rohe's definition that "Architecture is the will of an epoch translated into space."[14]

The quantifiable CH2 goal was to use efficiently the natural resources, hence to reduce waste and to save energy: according to the estimations, electricity consumption was reduced by 82%, gas by 87%, and water by 72%.

---

[14] From *Der Querschnitt,* 1924, in https://modernistarchitecture.wordpress.com/2010/10/25/ludwig-mies-van-der-rohe%E2%80%99s-%E2%80%9Carchitecture-and-the-times%E2%80%9D-1924/

The design process began with an interdisciplinary 2-week charrette[15] where all the partners were invited to participate (architects, engineers, artists, environmental experts, future occupants, the CSIRO, and the Sustainable Energy Authority of Victoria [29] and was followed by regular design sessions during the next 8 months. The idea was brilliant, as it gave the possibility of bridging gaps and addressing potential issues from the early stages of the project and throughout all the stages of the construction process (Fig. 5.25).

Fig. 5.25 Architect Mick Pearce. Council House 2, Melbourne, Australia. (Photo: Jonathan Lin CC BY-SA 2.0. Source: https://www.flickr.com/photos/jonolist/2653077336)

According to Mick Pearce,[16] "Every aspect of the building has been examined and rethought from first principles, evolving new precepts that are based in the desire to be as true as possible to the fundamental "laws of nature". The design

---

[15] Workshop oriented on design. The name derives from the cart—the charrette—that was collecting the work of the students of the École des Beaux-Arts in Paris, at the end of the term. Art and architecture students of the nineteenth century, working up to the deadline (nothing has changed in this respect), were sometimes climbing literally in the cart, for final touches: you lose the cart, you lose the deadline.

[16] Zimbabwean-born architect who designs using biomimicry principles; in doing so, the buildings have low maintenance and low running costs.

philosophy is concerned with developing appropriate architectural responses that are a direct and honest expression of the biodynamic relationships that nature uses in her own designs" [30].

These ideas are supported in the foreword of the ten technical studies carried out at the end of the construction phase and uploaded on the City of Melbourne site [31, Courtesy DesignInc]:

> *"From the revolutionary cooling storage system in the basement to vertical gardens and wind turbines on the roof, the building has sustainable technologies integrated throughout its 10 storeys. Although the majority of the technologies and principles adopted in the building are not new, never before in Australia have they been used in an office building in such a comprehensive and interrelated fashion. This includes innovations such as: using thermal mass for improving comfort; phase changing material to reduce peak energy demands and energy use; generating electricity onsite from natural gas; and using waste heat for cooling and heating."*

The emphasis was set on the peoples' needs so both the indoor and the outdoor environment (and the relation with the urban landscape) were carefully studied: indoor air quality (temperature, humidity, ventilation), lighting management (artificial and natural lighting including glare reduction), as well as the relaxing atmosphere created by plants and the rooftop garden were considered.

Taking into consideration Melbourne's climate,[17] most of the necessary heating is provided by the occupants and the working equipment. Temperature control relates mostly to the cooling of the building [32. Courtesy DesignInc].

Instead of using regular HVAC systems, displacement ventilation was adopted, with differentiated strategies for night and for day but with the same result: pushing the warmer, contaminated air upward and exhausting it through stacks. During the day, chilled panels placed on precast ceilings provide a permanent flow of descending cold air that, combined with the fresh air fed at the floor level, favors buoyancy and the evacuation of stale air [33. Courtesy DesignInc]. The low temperature of the chilled panels is resulting from the water that is circulated in the pipes during the day and cooled by phase changing materials (PCM) that in turn cool the beams and the prefabricated reinforced ceiling segments. Eventually, due to their thermal mass, these heavy building elements ensure a stable indoor temperature throughout the day.

Night purge allows exterior, cooler air to enter through the operable windows (Fig. 5.26, left), cooling the whole building.

---

[17] Located on the shores of Port Phillip Bay, Melbourne is situated on the 38th parallel South. Its climate is mild, temperate, with June and July the coldest (and windiest) months of the year. The mean temperatures are between 7 (in winter) and 25°C (in summer).

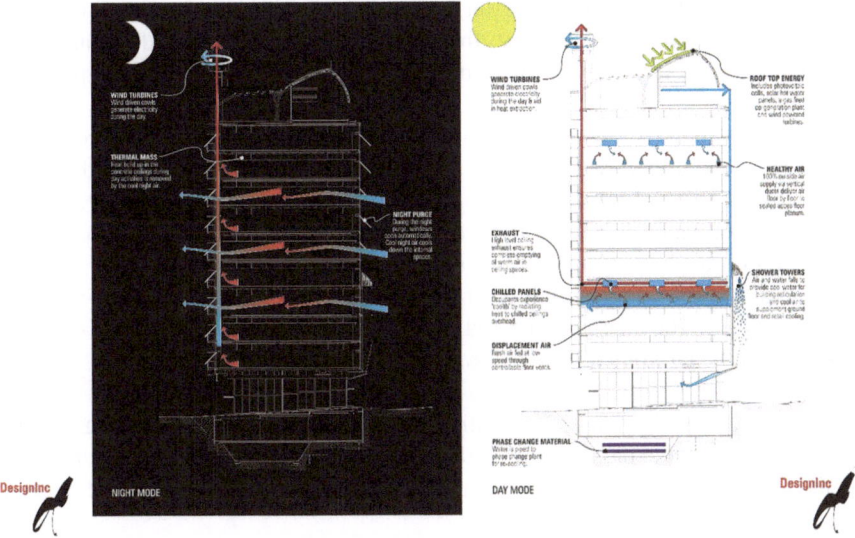

Fig. 5.26 Bioclimatic sections: night and day. (By courtesy of designInc)

The facades are responsive to the environment as well as to the necessities of the workers: the most exposed facade, North oriented (facing the equator), is shadowed by the planted balconies and protected by light shelves (Fig. 5.26, right); the Southern facade is provided with 17 m shower towers (Fig. 5.24, right) that humidify and cool the air at the street (retail) level; the Western glass facade has pivoting timber shutters (Fig. 5.24) that protect it, according to the daily movement of the sun.

Natural light enters through windows that are dimensioned according to the position on the facade: larger on the lower floors and smaller on the upper floors where no shading from the neighboring buildings affects the daylighting; light shelves and appropriate finishing surfaces direct the sunbeams and avoid glare [34. Courtesy DesignInc].

Considering the water issue of Melbourne, a strategy for water management was carried out and implemented. Thus, the vertical garden and the rooftop garden are irrigated (and subirrigated) with stored rainwater, treated grey- and blackwater, as well sewer mining; potable water is supplied by recirculating it from the sprinkler system [35. Courtesy DesignInc].

The surface of displaced vegetation at the ground level was compensated by a rooftop garden and the climbing vines on the northern facade.

The selection criteria of the building materials was made according to their recyclability: the wooden shutters are made of reclaimed timber and the metallic structure is 100% originating from recuperated steel. The whole philosophy of choosing the building products, components, and equipment is based on two main goals: to produce energy, to harvest and use it, as well as to provide a minimum (if at all) maintenance cost of the assemblies. Hence, plaster and paint were replaced by concrete, glass, metal, and wood surfaces that maintain their aspect without any interventions. In the same context, the possibility of relocation and discarding of structures and components with a minimum of waste was established from the design stage.

Active systems complete and enhance the passive measures: wind cowls use negative pressure to exhaust the hot, stale interior air that rises through the chimneys, solar PV and heating panels produce power that covers about 60% of the building's hot water demands, and a cogeneration plant provides heat, steam, and electricity for the heating and cooling system. Recovered energy was also taken into consideration: the energy produced by the spinning of the cowls (Figs. 5.23 and 5.24) as well as the energy generated when the elevators brake to stop is captured and used in the building's interior [36. courtesy DesignInc].

Efficient fixtures, light and water fittings, LCD displays, and other energy-saving devices complete the picture of a building where all the involved parties were aware of the fact that even the smallest step taken represents a contribution in the effort of saving energy.

Why were these buildings—and not others—chosen? Because the scientific approach that has always been the building design, carried out by interdisciplinary teams, led to innovative solutions that need to be understood, known, and used. After completion, these buildings were, for several years, under observation: specialists, theoreticians, and architects monitored their behavior and registered their performances. True, not all the expected levels of performance were met, not all the scenarios were validated by life,[18] but conclusions were drawn, and lessons were learned—lessons that can lead to better design, in respect to nature and with all the living entities that populate it. The innovative design is founded on ancient principles. In other words, on knowing and understanding the environment and the laws that govern it. The use of passive design strategies represents the simplest, safest, and least expensive way of saving energy. In all the examples that were presented in this final chapter, the architects turned to thermal mass, solar gain, vegetation, natural ventilation, and natural light—passive leverages—not because they are free of charge (which they are) but because they are healthy for the occupants. Technology came as a supporting tool, to provide the difference between what is supplied by passive means and what is needed.

The aim of this book—and chapter in particular—is not to avoid the use of technology and return to merely applying ancestral principles, on the contrary, but to remind architects and non-architect that

– Architecture is a complex transdisciplinary approach; it cannot exist without the collaboration of an increasing range of specialists as the technologies become ever-more diverse. The sooner the interdisciplinary team begins to work, the

---

[18]Which is normal, as nature has its own way and our models only simulate as close as possible what we know about it.

fewer compromises need to be made in the building process, and hence, better final outcome can be expected.

– Energy-efficient buildings were constructed since the Antiquity, with the means and limitations of each era. For several reasons, only some buildings passed the test of time, but what they prove is that the appropriate design that considers local conditions, the deep understanding of the building physics in shaping the space, and the use of adequate technologies may be the reasons for their survival.

And that a building is not just a shelter but a *dwelt space*, that bears *character* and personality.

References

1. Norberg-Schultz, C. (1979). *Genius loci. Towards a phenomenology of architecture*. Rizzoli.
2. Klepeis, N., Nelson, W., Ott, W., et al. (2001). The National Human Activity Pattern Survey (NHAPS): A resource for assessing exposure to environmental pollutants. *Journal of Exposure Science & Environmental Epidemiology, 11*, 231–252. https://doi.org/10.1038/sj.jea.7500165
3. https://corporate.steiff.com/en/company/history/
4. https://corporate.steiff.com/en/steiff-teddy/history/
5. http://www.historyofdolls.com/doll-history/history-of-ball-jointed-dolls/
6. https://vielfaltdermoderne.de/en/steiff-factory/
7. https://www.kycomfort.com/history-of-residential-air-conditioning/
8. Streicher, W., Heimrath, R., & Hengsberger, H., Mach, T., et al. BESTFAÇADE best practice for double skin façades EIE/04/135/S07.38652 WP 1 Report "State of the Art" Reporting period: 1.1.2005–31.12.2005, p.11.
9. https://www.casabatllo.es/en/antoni-gaudi/casa-batllo/history/
10. https://www.webpages.uidaho.edu/arch504ukgreenarch/casestudies/queensbldg-demontfortu.pdf
11. https://www.dmu.ac.uk/campus/venues/gallery/queens-building.aspx
12. Short, C. A. (2018, January). The recovery of natural environments in architecture: Delivering the recovery. *Journal of Building Engineering, 15*, 328. https://doi.org/10.1016/j.jobe.2017.11.014
13. https://ec.europa.eu/eurostat/statistics-explained/index.php?title=Glossary:Carbon_dioxide_emissions
14. https://www.fosterandpartners.com/projects/commerzbank-headquarters
15. Mary Pepchinski with its naturally ventilated skin and gardens in the sky, Foster and Partners' Commerzbank reinvent the skyscraper, Architectural Record 1998, January, pp. 69–79.
16. https://www.urbansystems.design/commerzbank-hq-frankfurt-germany
17. Banfi, M., & Guthrie, A. (1999). Kanak cultural centre, Noumea, New Caledonia, The Arup Journal , pp. 26–29. https://www.arup.com/globalassets/downloads/arup-journal/the-arup-journal-1999-issue-2.pdf
18. https://www.cepf.net/our-work/biodiversity-hotspots/new-caledonia
19. https://www.fondazionerenzopiano.org/media/project_document_en/5aec2a9e4da47757685879.pdf?v=v1.2
20. Twinn, C. (2003). BedZED, The Arup Journal , p. 11. https://www.arup.com/projects/the-arup-journal-2000s/the-arup-journal-2003-issue-1/
21. Schoon, N. (2016). *The BedZED story The UK's first large-scale, mixed-use eco-village*. https://storage.googleapis.com/www.bioregional.com/downloads/The-BedZED-Story_Bioregional_2017.pdf

22. https://www.neighbourhoodguidelines.org/zero-energy-ecovillage-bedzed-uk
23. Dabija, A.-M. (2021). *Alternative envelope components for energy efficient buildings.* Springer Nature.
24. Webb, S., & Downie, P. (2023, May). Revisit: BedZED in Beddington, UK by ZEDfactory. *Architectural   Record.*   https://www.architecturalrecord.com/ext/resources/archives/backissues/1998-01.pdf?883630800
25. Short, C. A., Lomas, K. J., & Woods, A. (2004). Design strategy for low-energy ventilation and cooling within an urban heat Island. *Building Research & Information, 32*(3), 187–206. https://doi.org/10.1080/09613210410001679875
26. Building Research & Information. (2004, May–June). *32*(3), (pp. 187–206). Taylor & Francis Ltd. http://www.tandf.co.uk/journals. https://doi.org/10.1080/09613210410001679875.
27. https://www.architectsjournal.co.uk/archive/specifiers-choiceschool-of-slavonic-studies-ucl
28. Webb, S. (2005, February). The integrated design process of CH2, environment design guide, CAS 36, p. 1., https://www.melbourne.vic.gov.au/SiteCollectionDocuments/ch2-case-study.pdf
29. https://www.melbourne.vic.gov.au/SiteCollectionDocuments/ch2-case-study.pdf
30. https://www.mickpearce.com/CH2.html
31. https://www.melbourne.vic.gov.au/SiteCollectionDocuments/ch2-nature-aesthetics-technical-paper.pdf.
32. Design Snap Shot 05 compiled by Dominique Hes, dhes@unimelb.edu.au from input by the entire design and construction team.
33. Snap shot compiled by Dominique Hes, dhes@unimelb.edu.au from input by the entire design and construction team.
34. https://www.melbourne.vic.gov.au/SiteCollectionDocuments/ch2-workplace-environment-study-outline.pdf
35. Design snap shot 09: Water Initiatives compiled by Dominique Hes, dhes@unimelb.edu.au from input by the entire design and construction team.
36. Green, J. (2021). *Good energy* (pp. 194–199). Princeton Architectural Press.

Index